Measuring the Impacts of Federal Investments in Research

A WORKSHOP SUMMARY

Steve Olson and Stephen Merrill, Rapporteurs

Committee on Measuring Economic and Other Returns on Federal Research Investments

Board on Science, Technology, and Economic Policy
Committee on Science, Engineering, and Public Policy

Policy and Global Affairs

THE NATIONAL ACADEMIES

THE NATIONAL ACADEMIES PRESS
Washington, D.C.
www.nap.edu

THE NATIONAL ACADEMIES PRESS   500 Fifth Street,   N.W. Washington, DC 20001

NOTICE: The project that is the subject of this report was approved by the Governing Board of the National Research Council, whose members are drawn from the councils of the National Academy of Sciences, the National Academy of Engineering, and the Institute of Medicine. The members of the committee responsible for the report were chosen for their special competences and with regard for appropriate balance.

This study was supported by Contract/Grant No. SMA-1019816 between the National Academy of Sciences and the National Science Foundation; Contract/Grant No. N01-OD-4-2139, TO #231, between the National Academy of Sciences and the National Institutes of Health; Contract/Grant No. G104P00159 between the National Academy of Sciences and the U.S. Geological Survey; Contract/Grant No. 59-9000-0-0093 between the National Academy of Sciences and the U.S. Department of Agriculture; Contract/Grant No. EP-11-H-001414 between the National Academy of Sciences and the Environmental Protection Agency; Contract/Grant No. DE-SC000614 between the National Academy of Sciences and the Department of Energy; Contract/Grant No. NNH10CC488,TO #5, between the National Academy of Sciences and NASA. Any opinions, findings, conclusions, or recommendations expressed in this publication are those of the author(s) and do not necessarily reflect the views of the organizations or agencies that provided support for the project

International Standard Book Number -13:978-0-309-21748-4
International Standard Book Number -10:0-309-21748-2

Additional copies of this report are available from the National Academies Press, 500 Fifth Street, N.W., Lockbox 285, Washington, DC 20055; (800) 624-6242 or (202) 334-3313 (in the Washington metropolitan area); Internet, http://www.nap.edu.

Cover: The cover design incorporates a feature of the 1924 National Academy of Sciences building in Washington. Sculpted by Lee Lawrie, the bronze cheneau, running the length of the roof, features alternating figures of owls and lynxes, symbolizing wisdom and alert observation, respectively.

Printed in the United States of America

THE NATIONAL ACADEMIES
Advisers to the Nation on Science, Engineering, and Medicine

The **National Academy of Sciences** is a private, nonprofit, self-perpetuating society of distinguished scholars engaged in scientific and engineering research, dedicated to the furtherance of science and technology and to their use for the general welfare. Upon the authority of the charter granted to it by the Congress in 1863, the Academy has a mandate that requires it to advise the federal government on scientific and technical matters. Dr. Ralph J. Cicerone is president of the National Academy of Sciences.

The **National Academy of Engineering** was established in 1964, under the charter of the National Academy of Sciences, as a parallel organization of outstanding engineers. It is autonomous in its administration and in the selection of its members, sharing with the National Academy of Sciences the responsibility for advising the federal government. The National Academy of Engineering also sponsors engineering programs aimed at meeting national needs, encourages education and research, and recognizes the superior achievements of engineers. Dr. Charles M. Vest is president of the National Academy of Engineering.

The **Institute of Medicine** was established in 1970 by the National Academy of Sciences to secure the services of eminent members of appropriate professions in the examination of policy matters pertaining to the health of the public. The Institute acts under the responsibility given to the National Academy of Sciences by its congressional charter to be an adviser to the federal government and, upon its own initiative, to identify issues of medical care, research, and education. Dr. Harvey V. Fineberg is president of the Institute of Medicine.

The **National Research Council** was organized by the National Academy of Sciences in 1916 to associate the broad community of science and technology with the Academy's purposes of furthering knowledge and advising the federal government. Functioning in accordance with general policies determined by the Academy, the Council has become the principal operating agency of both the National Academy of Sciences and the National Academy of Engineering in providing services to the government, the public, and the scientific and engineering communities. The Council is administered jointly by both Academies and the Institute of Medicine. Dr. Ralph J. Cicerone and Dr. Charles M. Vest are chair and vice chair, respectively, of the National Research Council.

www.national-academies.org

COMMITTEE ON MEASURING ECONOMIC AND OTHER RETURNS ON FEDERAL RESEARCH INVESTMENTS

NEAL LANE (Co-Chair), Malcolm Gillis University Professor, Rice University

BRONWYN HALL (Co-Chair), Professor of Economics, University of California at Berkeley and University of Maastricht

ALAN GARBER, Henry J. Kaiser, Jr. Professor and Professor of Medicine; Director, Center for Health Policy, Stanford University

PAULA STEPHAN, Professor of Economics, Georgia State University

PRABHU PINGALI, Deputy Director, Agricultural Development, Global Development Program, The Bill and Melinda Gates Foundation

WALTER POWELL, Professor of Education, Stanford University and External Professor, The Santa Fe Institute

DAVID GOLDSTON, Director, Government Affairs, Natural Resources Defense Council

ALEXANDER FRIEDMAN, Chief Investment Officer, UBS Wealth Management

JOHN STASKO, Professor and Associate Chair, School of Interactive Computing, Georgia Institute of Technology

ALFRED SPECTOR, Vice President of Research and Special Initiatives, Google, Inc.

ERIC WARD, President, The Two Blades Foundation

NEELA PATEL, Director of External Research, Global Pharmaceutical R and D, Abbott Laboratories

MICHAEL TURNER, Bruce V. and Diana M. Rauner Distinguished Service Professor, Kavli Institute for Cosmological Physics, The University of Chicago

Staff

STEPHEN A. MERRILL, Project Director

GURUPRASAD MADHAVAN, Program Officer and Project Co-director

KEVIN FINNERAN, Director, Committee on Science, Engineering, and Public Policy

STEVE OLSON, Consultant Writer
DANIEL MULLINS, Program Associate
CYNTHIA GETNER, Financial Associate

COMMITTEE ON SCIENCE, ENGINEERING, AND PUBLIC POLICY

National Academy of Sciences
National Academy of Engineering
Institute of Medicine

GEORGE WHITESIDES (Chair), Woodford L. and Ann A. Flowers University Professor, Harvard University

LINDA ABRIOLA, Dean of Engineering, Tufts University

CLAUDE CANIZARES, Vice President for Research, Associate Provost and Bruno Rossi Professor of Experimental Physics, Massachusetts Institute of Technology

MOSES CHAN, Evan Pugh Professor of Physics, Pennsylvania State University

RALPH CICERONE (Ex-Officio), President, National Academy of Sciences

PAUL CITRON, Retired Vice President, Technology Policy and Academic Relations, Medtronic, Inc.

RUTH DAVID, President and Chief Executive Officer, ANSER (Analytic Services), Inc.

HARVEY FINEBERG (Ex-Officio), President, Institute of Medicine

JUDITH KIMBLE, Investigator, Howard Hughes Medical Institute; Professor of Biochemistry and Molecular Biology and Medical Genetics, University of Wisconsin

DAN MOTE, Jr. (Ex-Officio), President and Glenn Martin Institute Professor of Engineering, University of Maryland

PERCY PIERRE, Vice President and Professor Emeritus, Michigan State University

ALBERT REECE, Vice President for Medical Affairs, Bowers Distinguished Professor and Dean, School of Medicine, University of Maryland, Baltimore

SUSAN SCRIMSHAW, President, The Sage Colleges

WILLIAM SPENCER, Chairman Emeritus, SEMATECH

MICHAEL TURNER, Bruce V. and Diana M. Rauner Distinguished Service Professor, Kavli Institute for Cosmological Physics, The University of Chicago

CHARLES VEST (Ex-Officio), President, National Academy of Engineering
NANCY WEXLER, Higgins Professor of Neuropsychology, Columbia University

Staff

KEVIN FINNERAN, Director
THOMAS ARRISON, Senior Program Officer
GURUPRASAD MADHAVAN, Program Officer
PETER HUNSBERGER, Financial Associate
MARION RAMSEY, Administrative Associate
NEERAJ GORKHALY, Research Associate

ACKNOWLEDGMENT OF REVIEWERS

This report has been reviewed in draft form by individuals chosen for their diverse perspectives and technical expertise, in accordance with procedures approved by the National Academies' Report Review Committee. The purpose of this independent review is to provide candid and critical comments that will assist the institution in making its published report as sound as possible and to ensure that the report meets institutional standards for quality and objectivity. The review comments and draft manuscript remain confidential to protect the integrity of the process.

We wish to thank the following individuals for their review of this report: George Bo-Linn, Gordon and Betty Moore Foundation; Susan Cozzens, Georgia Institute of Technology; Kenneth Gertz, University of Maryland; Diana Hicks, Georgia Institute of Technology; and Peter Hussey, RAND Corporation.

Although the reviewers listed above have provided many constructive comments and suggestions, they were not asked to endorse the content of the report, nor did they see the final draft before its release. Responsibility for the final content of this report rests entirely with the rapporteurs and the institution.

CONTENTS

1

INTRODUCTION AND OVERVIEW

The enactment of the America COMPETES Act in 2006 (and its reauthorization in 2010), the increase in research expenditures under the 2009 American Recovery and Reinvestment Act (ARRA), and President Obama's general emphasis on the contribution of science and technology to economic growth have all heightened interest in the role of scientific and engineering research in creating jobs, generating innovative technologies, spawning new industries, improving health, and producing other economic and societal benefits. Along with this interest has come a renewed emphasis on a question that has been asked for decades: Can the impacts and practical benefits of research to society be measured either quantitatively or qualitatively?

On April 18-19, 2011, the Board on Science, Technology, and Economic Policy (STEP) of the National Research Council and the Committee on Science, Engineering and Public Policy (COSEPUP), a joint unit of the National Academy of Sciences, the National Academy of Engineering, and the Institute of Medicine, held a workshop to examine this question The workshop brought together academic researchers, research and development (R and D) managers from private industry, representatives from government agencies, leaders of philanthropic organizations, and others to look at the very broad range of issues associated with evaluating the returns on federal investments (Appendix A). Speakers included researchers who have worked on the topic for decades and early-career researchers who are pioneering non-traditional approaches to the topic. In recent years, new research has appeared and new data sets have been created or are in development. Moreover, international interest in the topic has broadened substantially— in Latin America and Asia as well as in Europe. The workshop included presentations by speakers from abroad to gain their perspectives on methods of analysis. The workshop sought to assemble the range of work that has been done in measuring research outcomes and to provide a forum to discuss its methods. The workshop's goal was not to identify a single best method or few best methods of measuring research impacts. The workshop considered methodological differences across fields of research to identify which can be applied to the broad range of federal research funding. It did not address the role of federal funding in the development of technology.

The workshop was motivated by a 2009 letter from Congressman Rush Holt (D-New Jersey). He asked the National Academies to look into a variety of complex and interconnected issues, such as the short-term and long-term economic and non-economic impact of federal research funding, factors that determine whether federally funded research discoveries result in economic benefits, and quantification of the impacts of research on national security, the environment, health, education, public welfare, and decision making. "Discussing the economic benefits of research is critical when discussing research programs during the annual federal appropriations process," he wrote. Obviously, no single workshop could examine all of those questions, but it laid the groundwork for such an inquiry.

The workshop was sponsored by seven federal agencies: the National Science Foundation (NSF), the National Institutes of Health (NIH), the U.S. Department of Agriculture (USDA), the Environmental Protection Agency (EPA), the U.S. Geological Survey (USGS), the National Aeronautics and Space Administration (NASA), and the Department of Energy (DOE). It was organized by a planning committee co-chaired by Neal Lane, Malcolm Gillis University Professor at Rice University and former director of NSF and the Office of Science and Technology Policy (OSTP), and Bronwyn Hall, Professor at the University of California, Berkeley, and the University of Maastricht.

Consistent with Congressman Holt's concerns, the planning committee focused the workshop on broad social effects of public research investments – economic growth, productivity, and employment, social values such as environmental protection and food security, public goods such as national security, and the behavior of decision-makers and the public. The near-term outputs of research— scientific publications and other communications, citations to previous work, research collaborations and networks, and even patents resulting from R and D— were a not a principal focus of the meeting. Arguably, scientific and technical training is a near-term output of research but was featured in the workshop discussion because of its relationship to job creation and wage growth. Moreover, a large proportion of the technical professionals trained in research is subsequently employed in other than research occupations. The planning committee did not stipulate a timeline for the research impacts of interest, although policymakers' interest is concentrated on the short-to medium-term and the measurement challenge becomes greater the longer the time horizon.

This summary of the workshop provides the key observations and suggestions made by the speakers at the workshop and during the discussions that followed the formal presentations. The views contained in this summary are those of individual workshop participants and do not represent the views of workshop participants as a whole, the organizing committee, STEP, COSEPUP, or the National Academies. The summaries of the workshop discussions have been divided into eight chapters. After this introductory chapter, chapter 2 looks at several broad issues involved in the use of performance measures for research. Chapter 3 examines the direct impacts of research on the economy and the quality of life. Chapter 4 considers a closely related topic: the effects of biomedical research on health. Chapter 5 reviews other impacts of research that are not necessarily reflected in economic markets, including international development, agricultural advances, and national security. Chapter 6 moves on to what many speakers cited as one of the most important benefits of research: the training of early career scientific investigators who go on to apply their expertise and knowledge in industry, government, and academia. Chapter 7 summarizes the views of analysts from the United Kingdom, the European Union, and Brazil, highlighting the somewhat different approaches to similar problems being taken in other countries. Chapter 8 examines the emergence of new metrics that may be more powerful in assessing the effects of research on a wide variety of economic and societal indicators. And chapter 9 presents observations made during a final panel presentation on the pitfalls, progress, and opportunities offered by continuing work on measuring the impacts of federal investments in research.

Remarks of Congressman Rush Holt (D-NJ)

At the beginning of the workshop, Congressman Rush Holt, whose 2009 letter initiated the process leading to the workshop, addressed the group by video. His remarks have been slightly shortened.

I can't emphasize strongly enough the importance of your gathering. Measuring the impact of federal investments in research is a critical need for both government and society. We are living in what may become a pivotal time in our history. For well over half a century we have mined the investments that we made in the immediate aftermath of the Second World War and the fear that gripped us after the launch of Sputnik, from the airplane to the aerospace industry, and from the semiconductor to the Internet. American scientists have built the foundation of the strongest economy in the world.

But the Sputnik era is over. American leadership and our shared prosperity are in peril. As President Obama has said, we're in need of another Sputnik moment. According to the World Economic Forum's latest Global Competitiveness Report, the United States ranks fourth in global competitiveness behind Switzerland, Sweden, and Singapore. Further, the World Economic Forum ranks the United States forty-eighth in the quality of math and science education in our schools. Of course, any such rankings of competitiveness or economic or educational achievement are subject to challenge under methodology and, further, those rankings may not be measuring what really can make or keep the United States great or prosperous. However, today 77 percent of global firms planning to build new R and D facilities say they will build them in China or India, not in the United States. In 2009, 51 percent of U.S. patents were awarded to non-U.S. companies. China has gone from fifteenth place to fifth in international patents. Other countries are investing and implementing many of the changes suggested five years ago here in the United States while we continue to hedge and debate. We're losing our leadership position and our edge in the global economy.

History suggests that our long-term economic prosperity depends on maintaining a robust, modern innovation infrastructure and educational system. That's why some of us worked hard to include $22 billion in new R and D funding in the American Recovery and Reinvestment Act. Those funds were an important short—and long-term boost for our economy— short-term in hiring lab technicians and electricians to wire the labs and administrators and clerks to handle the programs, long-term in bringing innovations yet to be determined. Sustainable economic growth will require a sustained investment.

Although our economy has made progress, it continues to struggle. We're facing a time of serious budget pressure and, perhaps more serious, political pressure that could imperil the support and funding for federal research and development. Some people are suggesting significant cuts for agencies like NSF, NIST, DOE, NIH, NASA, and EPA.

We must be careful stewards of public funds. We need to ensure that our money is being used wisely and efficiently on programs that meet our objectives: creating jobs, building the economy, and creating a sustainable energy future, for example. Yet it is clear to me that cutting federal research funds is not a wise way to balance our budget.

Decision making, whether individual or Congressional, often happens through anecdotes. Nevertheless, we have to be intellectually honest. We have to make sure that the anecdotes are based on something substantial. We need data that will show us what is working and who is being put to work. Evidence can triumph over ideology—sometimes.

You are taking seriously the responsibility to provide hard facts and evidence about our investments. Together, you are building the infrastructure that we need to answer these important questions. I believe that our technological leadership and the foundation of our whole economy depend on it.

2

THE USES AND MISUSES OF PERFORMANCE MEASURES

Economists, policy analysts, and other scholars have studied the returns from federal research investments for decades, and they have made considerable progress. But basic questions still have only partial answers: What percentage of the gross domestic product (GDP) should be devoted to research and development? How should research dollars be allocated among fields of research? Which institutions and researchers can conduct research most efficiently and productively?

In the first session of the workshop, three speakers addressed the broad and complex issues that arise in attempts to answer these questions on the basis of empirical evidence. Each emphasized that the issues are exceedingly complex, and each offered a partly personal perspective on the workshop topic. Their observations and reflections provided a basis for many of the presentations that followed.

THE PROMISE AND THE LIMITS OF MEASURING THE IMPACT OF FEDERALLY SUPPORTED RESEARCH

The endeavor to measure the impacts of federally supported research has an inherent tension, said Irwin Feller, Senior Visiting Scientist at the American Association for the Advancement of Science (AAAS) and Professor Emeritus of Economics at Pennsylvania State University, who spoke on one of the two papers commissioned by the organizing committee in preparation for the workshop (Appendix C). One objective of performance measures is to guide public decision making. Yet the task can be so difficult—and sometimes counterproductive—that it leads to what Feller, quoting John Bunyan's *Pilgrim's Progress*, called the Slough of Despond. The basic problem, as

Einstein stated, is that "not everything that counts can be counted, and not everything that can be counted counts"—a phrase that was quoted several times during the workshop.

The Multiple Uses of Performance Measures

Performance measures have many uses, Feller continued. First, they are used to do retrospective assessments of realized, observed, and measured impacts. In this case, basic questions are: How has that program worked? Has it produced the results for which it was funded? How could these research advances contribute to societal objectives?

Second, performance measures can be used to assess the best direction in which to head. Is this where scientific advances will occur? Will these scientific advances lead to the achievement of societal objectives?

Finally, performance measures can benchmark accomplishments against historical or international measures and advocate for particular actions.

In each of these cases, performance measures have little relevance in the abstract, Feller said. They need to be related to the decisions at hand, and their promise and limitations depend on the decision being made. "They are quite necessary and productive for certain types of decisions, problematic for others, and harmful for others."

The context of performance measures determines much of their promise and limitations, according to Feller. A critical question is who is asking the questions. In a university setting, a promotion and tenure committee might ask about publications and citations while a dean or president might ask which areas of the university to support. In the federal government, a member of Congress might ask whether appropriations for a particular laboratory will produce jobs in his or her district, the director of OSTP might ask questions about recommendations to make to the President, and the director of the Office of Management and Budget (OMB) might ask about U.S. research expenditures relative to all other demands on the budget. Similarly, different federal agencies might ask different questions. NSF might want to know how to use research to advance the frontiers of knowledge, while the EPA might want to use science to support regulatory decisions.

Performance measures have been the focus of longstanding and diverse research traditions, Feller said. Over the course of four decades, he has studied patent data, bibliometrics, and many other measures related to research performance. The economics literature continues to produce more refined measures, better data, and new estimation techniques. Feller cited one study that used 37 performance measures in terms of outputs, outcomes, and impacts. Scorecards that compile measures, both nationally and internationally, also are proliferating. New theories, models, techniques, and datasets are producing an intellectual ferment in the use of performance measures. In addition, the community of practice is strengthening, which will increase the supply and use of research-based, policy-relevant performance measures. "This is a rich and fertile field for exploration, for discovery, and for development," Feller observed.

The Promise of Performance Measures

In terms of the promise of performance measures, they provide useful baselines for assessing several forms of accountability.

First, such measures provide evidence that an agency, laboratory, or individual is making good use of allocated funds.

Second, well-defined objectives and documentation of results facilitate communication with funders, performers, users, and others. Results become verifiable and quantifiable information on what has been done.

Performance measures focus attention on the ultimate objectives of public policy. Researchers and policymakers sometimes refer to the "black box" of innovation - the complex process of turning knowledge into applications - and much research done in economics and related disciplines tries to explain what goes on inside the black box.

Finally, performance measures can help policymakers avoid "fads" that direct attention in unproductive ways. Data can document that some phenomena do not have a solid evidentiary base and that it is time to move on.

The Limits of Performance Measures

An obvious limit on performance measures is that the returns on research are uncertain, long term, and circuitous. This makes it difficult to put research into a strict accountability regime. Doing so "loses sight of the dynamics of science and technology," Feller said.

In addition, impacts typically depend on complementary actions by entities other than the federal government. This is particularly the case as fundamental research moves toward technological innovation, implementation, and practice.

A less obvious limitation is that the benefits from failure are often underestimated by performance measures. Risk and uncertainty are inevitable in research, which means that research often generates negative results. Yet such results can redirect research into extremely productive directions, Feller said.

The selection of performance measure can also offer what Feller called a specious precision. Different measurable outcomes such as productivity, employment, competitiveness, and growth are not necessarily compatible with each other. There may also be tradeoffs among measures, so that greater accuracy in one generates greater uncertainty in the other.

The selection of performance measures can distort incentives. Research managers strive to improve performance on the measures selected, which can lead to results that are not necessarily compatible with longer-term objectives.

A final limitation, according to Feller, is that there is limited public evidence to date of the contributions that performance measurement has made to improve decision making.

Three Major Questions

Federal science policy must ask three big questions, Feller observed:
1. How much money should be allocated to federal research?
2. How much money should be spent across missions, agencies, or fields of research?
3. Which performers should conduct research, and what are the allocation criteria used to distribute these funds?

Performance measures do not provide a basis for answering the first of these questions. They do not indicate if the ratio of R and D to gross domestic product (GDP) should be 2.8 percent, 3 percent, 3.2 percent, 4 percent, or 6 percent. "I don't know if there is any evidence to support one level rather than the other," said Feller.

With regard to the allocation of money across fields, performance measures lead to multiple answers and therefore to multiple possible decisions. For example, bibliometric studies among journals might point toward the importance of biochemistry, economic research might point to the influence of computer engineering, and survey research on the use of scientific knowledge by industry might point to the need to support engineering and applied research fields. Of course, all scientific fields are connected to others, but that does not help make decisions about where to increase funding at the margin. "Depending on the methodology and the performance measures you use, you get different fields of science that tend to be emphasized," said Feller.

Performance measures have greater potential, Feller continued, in deciding among the performers of research, whether universities, government laboratories, non-governmental organizations, or other research institutes and among investigators. Agencies often have to make such decisions, along with decisions about the structure of research teams and centers. However, performance measures are currently underused for this purpose.

Do No Harm

It is critically important to "do no harm," Feller emphasized. A major goal of developing performance measures is to improve the quality of decision making. But there are dangers in relying too heavily on performance measures. For example, some states are discussing the use of performance measures to determine funding levels for higher education, despite their many limitations. Some policymakers "are moving pell-mell into the Slough of Despond, and I think that's what you want to avoid."

Policy analysts also must be careful not to overpromise what performance measures can do. Analysts will be called to account if their measures turn out to be mistaken and lead to harmful decisions, Feller concluded.

INNOVATION AS AN ECOSYSTEM

Daniel Sarewitz, Professor of Science and Society at Arizona State University, reinforced and expanded on Feller's comments. The fundamental assumption of the workshop, he said, is that federal investments in research have returns to society that can be measured. However, this assumption raises the much larger question of how the innovation system operates. Policymakers have a tendency to simplify the operation of the system. For example, they may draw a straightforward connection between basic research and applications and imply that the basic task is to speed the movement from the former to the latter. It is "discouraging," said Sarewitz, that policymakers still feel a need to present such simplifications to garner public support.

Rather than introducing performance metrics into an oversimplified narrative, Sarewitz continued, perhaps it would be better to improve the narrative. This requires re-examining the role of research in the broader innovation process.

The Features of Complex Systems

Case studies of the role of research in innovation reveal an extremely complex process in which research is an important element of the process but not the only important element. "Everything is connected to everything else," said Sarewitz. "It's an ecosystem, and all things flow in different ways at different times depending on who is looking when and where in the process." For example, technology often enables basic science to address new questions. Similarly, tacit knowledge acquired through the day-to-day practice of, for example, engineers or physicians can raise important questions for researchers. As an example, Sarewitz cited a statement by former NIH Director Harold Varmus that some cancer treatments are "unreasonably effective" but that it is hard to fund research on these treatments because such research is considered high risk. "I was stunned by this, because my view of the complexity of the innovation system is that if we understand that technologies and practices themselves are sources of problems that research can address, then one ought to see unreasonably effective cancer treatments as an incredibly potent attractor of research." However, the predominant model of research pursued at NIH is to understand the fundamental dynamics of a disease, which then will lead rationally toward the best treatments to use.

There is a deeper problem, said Sarewitz. In a complex system such as the innovation ecosystem, there is no reason to believe that optimizing the performance of any one part of the system will optimize or even necessarily improve the performance of the system as a whole. "Another way to put this is that research is not an independent variable in the innovation system. We generally don't know what the independent variables are. For analytical purposes there may not be any."

The connections that link the elements of the innovation system represent contextual factors that can be crucial determinants of performance. Factors such as trust among the people in an institution, administrative structures that allow for rapid learning and adaptation, or historical ties between different institutions that allow them to work together can be very important for determining the dynamics and ultimate success of complex innovation processes. These sorts of internal systems dynamics can be teased out through careful case studies, Sarewitz said. But they are very difficult to capture in de-contextualized and rigid performance measures.

The Policy Perspective

Policymakers have an array of tools that they can use to try to influence the behavior of complex innovation processes. However, just a few of these tools relate directly to research, and the relations among these tools are poorly understood. For example, analysts would have difficulty measuring and comparing the performance of intramural laboratories and extramural university research without also knowing the institutional contexts of the research performers.

More generally, research performance measures may reveal little about the value and contextual appropriateness of the full array of science policy tools. For example, tools like demonstration and procurement, especially as done by the Department of Defense, have been enormous drivers of innovation in the past, yet they are outside the domain of research performance measures. Given the importance of other factors, optimizing research performance could lead to undesired outcomes.

These undesired outcomes may even have ethical and moral dimensions, said Sarewitz. For example, policy decisions in the early 1980s accelerated the privatization of the results of publicly funded research and helped to elevate the importance of patents as an apparent indicator of innovation. However, these policy decisions have consequences that bear on equity to access of some of the products of publicly funded research. In the medical arena, to cite an example Sarewitz mentioned, they could have slowed innovation in socially important domains of research, such as the development of agricultural biotechnologies for developing countries.

Innovative Approaches

The science and technology policy and research communities have to engage as imaginatively as possible in expanding the array of approaches used to understand, assess, and talk about innovation processes and their outcomes in society, Sarewitz said. First, new understandings of complex innovation processes can be used to help improve policy making. Case studies, for example, can produce synthetic systems-oriented insights that can have a powerful and enriching impact on policy making and "hopefully, change the narrative."

Second, the science policy research community can do a better job of coming up with diverse performance criteria and measures that can support rather than displace qualitative insights. An interesting recent example involved the public policy analogues of market failures, which could be used to drive public investments in the same way that market failures have in the past (Bozeman and Sarewitz, 2005). "We don't know yet if this particular approach is going to turn out to be a valuable tool," said Sarewitz. "The point I'm trying to make is that the narrow array of things we are now measuring as indicators of performance of the innovation system, mostly matters of research productivity, is impoverished and we can and should do better."

Research is crucially important in innovation, Sarewitz concluded. But its importance is contextual and contingent in space, among institutions, and over time. "If decision makers focus on optimizing performance and the innovation enterprise based on measures that largely deal with research, research performance, and research outputs, they'll likely fail to achieve the goals that the public expects from the nation's R and D investment."

OVERCOMING THE CHALLENGES OF RESEARCH MEASURES

In a commentary on Feller's and Sarewitz's presentations, Alfred Spector, Vice President at Google, agreed that mechanisms are needed to determine the right amount, the proper balance, and the overall effectiveness of research investments. But he also pointed out that these mechanisms face several challenges.

First, measurement imposes overhead on the research community. Especially when the measurements do not seem to be related to specific outcomes, researchers can chafe at the time and effort involved in filling out forms or answering questions. If measurements were simple, overhead would be reduced. But the innovation system is complex and single measures can be misleading, which means that multiple measures are needed.

The act of measuring also can perturb the research being done. Spector cited an example from computer science involving the relative emphasis on patenting. He said that most people working in his field would conclude that greater emphasis on patenting would reduce the rate of innovation. "Most faculty agree that patents in computer science basically are almost always a bar that reduces the rate of innovation by creating rigidities and without the benefits of the economic incentives that are supposedly being provided. This may not be true in the biotechnologies, but it is true, I believe, in my field."

Some measures also may be outdated. For example, publications have been important in the past. But in computer science today, an important product of research is open source software that is broadly disseminated. Such dissemination is a form of publication, but it is not a refereed publication that traditionally has factored into evaluations. Similarly, open standards can be incredibly valuable and powerful, as can proprietary products that establish the state of the art and motivate competition.

Accounting for Overlooked Measures

Greater transparency can help overcome these challenges, said Spector. The growth of modern communication technologies makes transparency much more feasible today than in the past, providing a more open view of research outcomes. Similarly, better visualizations can produce representations that are useful to policymakers and the public in assessing the value of research.

One of the most important products of research, though it is sometimes overlooked, is the training of people, Spector said. "If you talk to most of my peers in industry, what we really care about as much as anything else is the immense amount of training that goes on through the research that's done." For example, venture capitalists would rate talent as the most important input into innovation.

Also, the diversity of research approaches can be an important factor in research. In computer science, for example, funding has come not only from the NSF, in which peer review largely determines what science will be done, but also from the Defense Advanced Research Projects Agency, which has a much more mission-oriented approach. "DARPA has made huge bets, primarily on teams that they believed would win those bets. That has also resulted in huge results." However research is measured, it has to accommodate different approaches to realize the advantages of diversity, Spector said.

Failure is an important aspect of research. If there is no failure in research projects, then they are not at the right point on the risk-reward spectrum, said Spector. Rewarding failure may not seem like a good thing, but for research it can be essential. At Google, said Spector, "we view it as a badge of honor to agree that a certain line of advanced technology or research is not working and to stop and do something else. I think we need to have measurements like that in the world at large, although it's clearly a challenging thing to do."

Finally, the potential for serendipity needs to be rewarded. "If everything is so strongly controlled, I have a feeling we'll do whatever the establishment feels is right and serendipity will be removed." Serendipity often produces the creative disruption that reshapes entire industries, Spector concluded.

DISCUSSION

In response to a question about using measures of research outcomes to increase commercialization, Feller warned against the distortions such initiatives can produce in agencies such as NSF. He agreed with Spector that industry is more interested in the trained students research produces than in specific findings or patents. Also, researchers are usually not able to predict with certainty the commercial or societal implications of their research.

However, Feller added that it may be possible to document the need for transformative research. For example, NSF has been funding Science and Technology Centers that are focused on emerging scientific opportunities with important societal implications, such as hydrological research or the atmospheric sciences, that can have difficulty obtaining funding through conventional channels because they are too risky or large. These centers can even be evaluated in part using traditional measures, such as the number of collaborators from different disciplines on papers. Sarewitz agreed that the agencies need to emphasize high-risk research because universities tend to pursue incremental change.

A workshop participant asked about the best way to evaluate research across an entire agency such as NSF to make decisions about the allocation of funding. Feller emphasized the importance of truth and transparency. He praised the work of the Science of Science and

Innovation Policy (SciSIP) Program at NSF and said that NSF needs to draw on the expertise being developed by the program and elsewhere in the agency. He also noted the need to re-fashion the Government Performance and Results Act (GPRA) to be more suited to research. At the same time, he noted the potential problem of researcher overhead and the need for measures to produce useful information. Sarewitz added that increments of information tend to have no impact on institutional decision-making processes.

Measures of research performance can help agencies "get their house in order," said Feller, since many allocation decisions are still internal to agencies. However, measures demonstrating positive research outcomes do not necessarily guarantee that Congress will continue to allocate funds for those programs. "At some point, these remain fundamentally political decisions with a strong tang of ideology," said Feller. Congress or OMB can always question, for example, whether a given program is an appropriate role for government.

Sarewitz pointed out that oversimplified narratives of innovation can contribute to this politization. If policymakers had a more sophisticated perspective on innovation, they would be more willing to accept a multi-faceted government role rather than devoting money solely to research. Spector added that information technologies provide new ways to disseminate these more sophisticated narratives, regardless of the origins and targets of those narratives.

David Goldston, who was on the planning committee for the workshop, pointed out that research funding decisions are inherently political. Showing that a given program is working usually answers a different set of questions than the opponents of a program are asking. Feller responded that dealing with the objections raised by the opponents of a program is like dealing with counterfactual scenarios, in which new scenarios can constantly be created that either have not been tested or are impossible to test. Nevertheless, the perspectives of policymakers on research have changed dramatically over the last few decades, so that they generally accept the need for the federal government to support fundamental research.

3

IMPACTS ON THE U.S. ECONOMY AND QUALITY OF LIFE

What is known about the contribution of research to GDP, productivity, wages, employment, and private sector R and D? Is there a basis for setting a target for aggregate research spending? How can the flow of knowledge from research into particular economic activities be measured? These were some of the questions addressed during the session of the workshop on the direct economic benefits of research spending. Three speakers looked at such issues as R and D's influence on productivity gains, the association between research activity and local labor market conditions, and citations in industrial patents to publicly funded research. These have been the principal avenues for measuring economic benefits of research

FEDERAL RESEARCH AND PRODUCTIVITY

From the 1950s to the 1970s, many studies examined the broad outcomes of federal R and D, but fewer studies have occurred in recent decades, said Carol Corrado, Senior Advisor and Research Director in Economics at the Conference Board. She presented recent results from investigations of the relationship between R and D and productivity, taking mostly a "30,000-foot perspective." She also emphasized a key prospective change in the U.S. national accounts. Starting in 2013, R and D spending will be capitalized as an investment instead of being treated, as it is now and has been historically, as an intermediate expense. This means that both private and public R and D will raise bottom-line GDP and national saving.

According to Corrado, the total U.S. R and D investment level has been stable since the 1980s as a share of GDP. Since 1959, the share of all R and D investment funded by the public sector has declined relative to that funded by the private sector, with rough stability in both sectors since about 2001. The total nominal R and D investment in 2007 was $407.5 billion, with business at $269.6 billion, government at $117 billion, universities at $10.6 billion, and nonprofits at $8.4 billion.

Corrado investigated the R and D intensity of eight industries over two time periods: the 1990s and the 2000s. When the R and D intensity of each industry matched Total Factor Productivity (TFP) estimates, as it did for the 1990s, R and D can be interpreted as the sole driver of productivity gains. The 1990s data also show that the computer industry, which was heavily subsidized by federal R and D, outperformed the others. In fact this industry seemed so exceptional that Corrado removed it to look solely at the other seven industries for more general trends. But even excepting computers, R and D appeared to be the sole driver of the productivity gains of the 1990s.

However, the same comparison showed that R and D contributed only 30 percent to the average industry productivity gain in the 2000s, Corrado said. This analysis had too little data to draw firm conclusions, according to Corrado. The analysis also was not able to measure the impacts of investments in the life sciences on human health, though the Bureau of Economic Analyses (BEA) is working to introduce a healthcare satellite account. Also excluded from this analysis was educational services, which may require a geographically localized approach.

The productivity growth of the 1990s suggests that the Internet and demand for networked devices were key drivers of economic activity in that decade, said Corrado. Government played "a classic role" in supporting new technology when several private companies worked with NSF to set up the first T1 telephone data line in 1987. This federal R and D created infrastructure and also helped to close "valleys of death" in the commercialization of research.

Corrado also called attention to the dwindling share of manufacturing in the U.S. economy. What does it mean for policy if the United States moves to an economy characterized by "designed in California, made in China"? she asked.

Finally, she observed that innovation is "more than science." Studies suggest that firms innovate based on intangibles such as product design, new business processes, and staff knowledge building, not just new research results. An estimate for 2001 put R and D's share of spending on all of these intangibles at just 16 percent, although R and D dollars could influence the outcome of spending on other intangibles.

Corrado said that the source of innovations needs to be better understood. For example, Virgin Atlantic holds a patent on the design of its first class cabins, which is one example of how the notion of a science and innovation policy can be broadened. The role of diffusion, which could help explain the changes from the 1990s to 2000s in the industries she analyzed, also needs more intensive study.

INDIRECT ECONOMIC BENEFITS OF RESEARCH

Government research expenditures are increasingly justified in terms of economic benefits such as job creation. But the practical benefits of research are disputed even by some scientists, said Bruce Weinberg, Professor of Economics and Public Administration at Ohio State University, and there is little accepted methodology for estimating these benefits.

Weinberg focused on "indirect benefits." He described these as the "productivity spillover benefits" beyond particular products or processes that develop out of research. Examples include a better trained workforce that generates higher productivity, solutions to industrial problems, new infrastructure, or hubs for innovation. Even if these spillover benefits turn out to be smaller than the direct benefits, "they are important and are increasingly driving the discussion about the cost and benefits of research."

One way to estimate the economic benefits of research is through job creation, but Weinberg noted that "this poses deep fundamental and practical problems." For example, if a job pays $50,000 a year, the value of the job to a person is really that amount minus what a jobholder would have been earning on another job. Also, as wages go up in science jobs, people may move to science from other occupations, which moves jobs from one sector to another rather than creating jobs.

Instead, Weinberg suggested focusing on outcomes—wages or productivity— in places where more science and research is carried out. What should be estimated, he said, is whether research leads to more productive industries in local economies.

Weinberg related measurements of research in particular cities to economic metrics of those cities. He asked whether wages and employment are better in cities where more research is being done. He also looked at measures of innovation such as patenting in cities with more science.

Based on preliminary results for U.S. metropolitan areas, a positive correlation exists between wages, employment, and academic R and D, he said. The results indicate that a 1 percent increase in academic R and D is associated with roughly 120,000 more people employed and $3 billion more earnings in a metropolitan area. Weinberg cautioned, however, that these results are far from definitive because of confounding factors. For example, science-intensive cities may be different from other cities, or workers may have different abilities across cities. "The literature hasn't really addressed the underlying challenges convincingly," he said.

"If I were to summarize the literature, I would say there is some evidence that science or research impacts wages, industrial composition, and patenting, but these estimates are weak," Weinberg concluded. For the future, it is important to think about productivity spillovers not simply in terms of job creation but by doing studies that "unpack the mechanisms by which science and research impact economic outcomes."

BEYOND CITATIONS AND PATENT REFERENCE COUNTS

A common way to measure knowledge flows among universities, government laboratories, and firms is through citations in patents to patent references (PR) assigned to universities, federal laboratories, or research institutes and citations to non-patent references (NPR) with an author affiliated with a university, federal laboratory, or research institute. Such references provide "rich data that can be used across industries and firms and over time," said Michael Roach, Assistant Professor of Strategy and Entrepreneurship at the Kenan-Flagler Business School at the University of North Carolina.

However, patent citations also suffer from some limitations, Roach acknowledged. Not all inventions are patented or even patentable, so such studies are limited in what they can observe. Similarly, not all knowledge flows are citable or cited. Firms may not want to disclose important developments, or industrial authors may overuse citations, which is a trend Roach has found in his research. As a result, citations likely mismeasure knowledge flows, either randomly or with a systematic bias.

In particular, NPR citations capture knowledge flows through channels of open science (such as publications), direct use of technological opportunities in new R and D projects, and knowledge flows to firms' applied research. NPR citations do not but should capture knowledge flows through contract-based relationships, intermediate use in existing projects, and knowledge flows to firms' basic research activities. All things considered, Roach concluded that citations likely understate the impact of public research on firms' performance.

Roach described a study done with Wesley Cohen (Roach and Cohen, 2011) that used the Carnegie Mellon R and D Survey of manufacturing firms to measure a firm's use of public research. The "key takeaway," according to Roach, was his calculation showing that the unobserved contribution of public research to innovative performance is comparable to what is observed. They estimate that observed knowledge flows account for about 17 percent of firms' innovative performance while unobserved flows account for about 16 percent.

Future research should concentrate on NPRs, Roach said. Though such data are costly to obtain, they are one of the best measures available to measure knowledge flows. He suggested that the National Bureau of Economic Research and the U.S. Patent and Trademark Office make NPR data more readily available to scholars.

Other external data could be used to measure knowledge flows, such as NSF's recently expanded Business R and D and Innovation Survey (BRDIS). Also, the origins of citations need to be better understood. "We need to be looking at the micro level," Roach said, echoing points made in the previous panel. Research needs to look at inventors, scientists, and firms— "trying to get inside that black box."

DISCUSSION

Alfred Spector of Google commented on Corrado's description of the change in national accounts making R and D a capital investment. Spector noted that firms currently expense research because they do not know what the results of the research will be. Corrado replied that while some business accountants are resisting the change, those who favor it say it can provide a "holistic picture of how and where firms make their investments. . . . What you set aside today to generate future consumption— in other words, what you forego today— is your investment." She explained that national accounts do not have to line up with firms' accounting practices.

The session moderator, Bronwyn Hall, said that publicly held firms use Financial Accounting Standards Board (FASB) policy for expensing R and D. An advantage is that expensing R and D offsets current income. The problem from an economic analysis perspective, Hall said, is that "in the United States, the value of firms even when the market is down is substantially higher than the value of their tangible capital assets." When one looks for what explains the difference, "capitalized R and D is the first thing" one sees.

In response to a question about how research funders can generate more positive spillover effects from research, Weinberg pointed out that research funding is more likely to have positive effects in nearby location than distant locations. Improvements in dissemination could enhance information flows, and there are many ways to study the impacts of this dissemination.

4

IMPACTS ON BIOMEDICAL AND HEALTH RESEARCH

The impacts of research on the health of people in the United States and around the world may not be measured by economic analyses, but historically these impacts have been among the most important benefits of research. Five speakers with diverse backgrounds addressed this topic at the workshop. They found evidence of substantial benefits while also identifying areas where benefits may be overlooked by current approaches. In addition, they called attention to problems with the funding of federal research, such as the damage up-and-down funding can do to the careers of researchers and the difficulty of allocating limited funds across categories of research.

REVIEWING THE LITERATURE ON HEALTH IMPACTS

Bhaven Sampat, Assistant Professor of Public Health at Columbia University, presented a brief summary of a commissioned paper (Appendix D) that discusses representative studies of the effects of publicly funded biomedical research on a range of outcomes. Public funding accounts for about one-third of all biomedical and health research, with NIH-sponsored research accounting for most of the federal component along with additional investments by NSF, DOE, DOD, USDA, and other agencies. In 2007, funding for biomedical research totaled slightly more than $100 billion.

Sampat showed a stylized albeit simplified view of the innovation system in which publicly funded R and D leads to improvements and efficiencies in the private sector, to new drugs and devices, and ideally to improved health outcomes (see Appendix D, Figure D-1). This flow of knowledge occurs through many channels. One channel encompasses publications, conference presentations, markets, and informal networks. A second channel is through the creation of prototypes for drugs and devices. Since the Bayh-Dole Act of 1980, these prototypes have tended to be developed in universities and licensed out to firms to turn them into successful products. A third channel includes funding for clinical trials and clinical research that informs clinical practice— such as the knowledge that doctors should give people an aspirin after a heart attack — along with funding of other applications-oriented work, such as contracts to fund the development of technologies and to conduct consensus conferences.

Sampat called attention to another impact of new biomedical technologies that is being discussed among health policy researchers. Most economists believe that biomedical technologies are the biggest source of long-run increases in health care costs. The clinical value from these technologies may exceed their costs, but technology-driven cost increases may be unsustainable, Sampat observed.

The Case of Cardiovascular Disease

Sampat described some of the literature on improvements in health outcomes that can be traced to research. Cutler and Kadiyala (2007) looked at improvements in cardiovascular disease mortality over the five decades beginning in 1950, when mortality fell by two-thirds. They concluded that about one-third of the advance is attributable to new high-technology treatments, one-third to new drugs, and one-third to behavioral changes such as not smoking and not eating salty or fatty foods. Using a standard evaluation of $100,000 per year of life used by health economists, they then computed the rate of return on investments in treatments. New treatments provided a 4-to-1 rate of return, while new behavioral knowledge produced a 30-to-1 rate of return. According to this paper, Sampat said, "the publicly funded R in R and D has been worth it."

This paper makes little mention of NIH or public research except for NIH's sponsorship of large epidemiological trials and conferences, which makes it hard to trace outcomes back to basic research. Another issue, said Sampat, is the counterfactual: What would have happened in cardiovascular disease absent any public funding in that area?

A paper by Heidenreich and McClellan (2007) focused on improvements in heart attack care. These authors go farther than Cutler and Kadiyala in relating changes in clinical practice to specific outputs of

R and D. The authors concluded that the medical treatments studied in clinical trials accounted for much of the improvement in heart attack outcomes. The challenges with this paper include the fact that the authors generally did not trace changes in clinical practice back to basic research. Also, clinical practice often leads to publicly funded R and D because informal learning by clinicians generates important research questions. This learning is often subsidized by Medicare payments to teaching hospitals and other non-research sources. Finally, clinical trials can lead to negative results and lead clinicians to stop doing things they were doing, which can be an unmeasured benefit to research.

Other Disease Categories

A statistical study by Manton et al. (2009) related mortality rates in four disease areas to lagged NIH funding for the relevant institutes from 1954 to 2004. For two of the diseases studied—heart disease and stroke — the authors found a relationship between funding and outcomes. For the two other diseases— cancer and diabetes - the evidence was weaker. But relying on funding aggregated by institute is difficult, as an institute can fund widely varying research. Also the counterfactual is hard to demonstrate since many factors could be driving changes in disease rates.

Over this time period, competing risks changed. One reason for the absence of a decline in cancer mortality— and maybe even an increase— is that fewer people are dying of heart disease, so they live longer and are more likely to develop cancer.

Relationship of Public and Private R and D

Papers by Toole (2007) and by Ward and Dranove (1995) sought to relate public sector R and D to private sector R and D and found strong evidence that they are complements rather than substitutes, in that public research tends to spur private research. Private and public R and D in a given area could be driven by scientific opportunity. For both forms of research, there are challenges linking R and D to health outcomes, Sampat observed.

Another line of research regarding private sector R and D is whether proximity to public sector scientists makes firms more productive. A range of studies have indicated that the answer is probably "yes," Sampat said, especially survey research asking firm R and D managers how much they rely on public sector R and D.

Sampat noted that in surveys the drug industry reports greater reliance on public sector R and D than do other industries. In contrast, the device industry tends to be at or below the mean in terms of reliance on public sector R and D. The drug industry relies mostly on medicine, biology and chemistry. The device industry relies on medicine and biology and, third, on materials science, which tends to be funded by NSF and DOD.

Sampat then turned to drug and device innovation. Very recent studies have used accounting methodologies to look at, for example, the impact of public sector R and D in producing drugs that are then marketed. In a study of drugs in FDA's Orange Book, about 10 percent of marketed drugs come from universities or public laboratories, meaning that these institutions hold key patents (Sampat and Lichtenberg, 2011). The number is higher, about 20 percent, for clinically important drugs.

The Case of HIV Drugs and Vaccines

HIV is a special case, Sampat observed. The role of the public sector in directly generating new drugs is much higher in HIV than in other arenas; nearly one third of drugs in this area rely on public sector research. Also, nearly all commercially and therapeutically important vaccines over the last 25 years have come from the public sector, according to Stevens et al. (2011). Surprisingly, efforts to relate funding by disease area to later drug innovation tend not to show much of an effect.

Device Development

In the areas of devices, Sampat described a case study by Morlacchi and Nelson (2011) on the development of the left-ventricular assist device (LVAD). The scientific understanding of heart failure remained quite weak throughout the period that the LVAD was developed. But NIH was holding consensus conferences to diffuse best practice and contracting with firms for device development and clinical trials. "The more applied side of the activities seems to be important" in this case, Sampat concluded.

Conclusion

The literature shows "consistent evidence of public sector funding on private sector innovative effort," Sampat concluded. The literature also shows that public sector R and D has been important in the generation of a non-trivial number of important drugs. However, it shows less impact on other innovative outputs.

"There is surprisingly little research on the health benefits of public sector biomedical R and D," Sampat observed. Most of the evidence to date is from the cardiovascular area. In addition, case studies point to the importance of public clinical research, applied research, and diffusion activities. Devices have important differences from drugs. And despite a good deal of discussion, there has not been much study of the effects of public sector research on health costs.

THE VOLATILITY OF FEDERAL R AND D SUPPORT

Richard Freeman, Herbert Ascherman Chair in Economics at Harvard University, addressed the unintended effects of variability in federal government funding for R and D. Using changes in the budgets of the National Institutes of Health as an example, Freeman said that chief among these effects is the damage done to people's careers by changes in grant rejection rates and increased uncertainty about future career prospects. Scientific careers "looked dicey" even after the Wall Street implosion and lay-offs in banking and consulting made finance less attractive.

Funding variability may also affect the productivity of scientific research. His study of the recent doubling of the NIH budget found that before the doubling period more papers were produced per dollar of grant than when more money was available. This decline in marginal productivity may make it easier to cut future funding due to "failing to meet 'promises,'" Freeman stated. By contrast, the private sector has not been as variable in its R and D support.

Gaps in Monitoring Science

Finally, Freeman suggested that the scientific community needs to do a better job of monitoring the state of science. For example, non-traditional measures of the supply of jobs might include real-time data from Internet job boards, searches for information about science and engineering jobs, and databases on Ph.D. dissertations. Downloads of working papers could indicate hot areas of research. Online science and social discussion groups and web-based communications from meetings and conferences could contain information useful to the policy community. Companies and other institutions should be accessing these databases regularly, he said.

Information on what industry is doing is weak. The aggregate amounts of money spent do little to map the steps to innovation. Even the NSF BRDIS survey provides little data beyond the amounts of money spent. Further, basic and applied research tend to be artificially divided, but anything that is an innovation is going to go back and forth between the two categories of research, Freeman observed.

MEDICAL DEVICE INNOVATION

Many people assume that the development of biomedical devices is similar to drugs, said Paul Citron, retired Vice-President at Medtronic, Inc., and now at the University of California, San Diego; but in fact "they have very different characteristics as they traverse the pathway from bench to bedside." Drugs tend to be more discovery-based and derived from in-house activity. Devices are engineering-based. A specification is generated, along with an idea of how to realize that specification. Moreover, devices evolve over time. The first device is very different from subsequent generations, whereas a drug tends to be static for its lifetime.

For devices, the timelines are longer and the markets are smaller than in the pharmaceutical industry, Citron explained. It is very rare for a medical device to have a billion dollar market, unlike pharmaceuticals.

The evolution of a device can be heavily influenced by federally funded research, according to Citron. For example, research can enable an industry to bring a device from concept to clinic. Federal funding can build the underlying knowledge needed to make a technology safe and effective. Federal research also can yield new materials, whereas the complexity and cost of coming up with a new biomaterial to be implanted in the human body can be beyond the ability of any one company. Clinical trials may be crucial in improving a device.

A successful outcome for interventions using a medical device depends on rigorous manufacturing, which can be improved through R and D cycles involving federal research. Most medical product recalls are due to manufacturing issues that arise after approval, which can be reduced through R and D.

Vivariums at academic centers are another crucial investment underwritten by federal support. Prototype products are often tested at these institutions, and even large companies may need to use academic centers for access to animals.

Finally, "probably the most important output of federal inputs," said Citron, is students. "We hire the products of the campus" because "that is where the intellectual horsepower for tomorrow resides."

Citron listed four criteria an industry uses to decide whether to pursue a project. (1) Does the technology fit with a company's internal capabilities? (2) Is the fit with the customer good? (3) What is the market opportunity, including the number of customers, price, and the details of application? (4) Finally, what is the time to market, including the time needed to satisfy the regulatory process?

MAKING DECISIONS IN THE PHARMACEUTICAL INDUSTRY

The pharmaceutical industry and regulatory bodies need to evaluate drugs thoroughly and expeditiously as they go through years of clinical development before gaining approval for use in the treatment of a particular disease state, observed Dr. Garry Neil, Corporate Vice President for Science and Technology at Johnson and Johnson. Dr. Neil's company discovers and develops therapeutic products and technologies that are evaluated by regulatory agencies around the world to assess the efficacy and safety of a product for its intended use. "We have set the bar very high for ourselves [about] what is expected and what we need to deliver to our stakeholders, and we take that very seriously."

Drug discovery spans years of study or phases of study, from prediscovery to post-marketing surveillance, which allows for continued follow up in a real-world setting after a therapy had been approved – but getting to the point of approval can be challenging. Typically thousands of compounds are synthesized to yield just a few potential candidates that enter preclinical study, and for every five thousand to ten thousand synthesized compounds, one approved drug on average may emerge. And for many reasons, the costs to develop new drugs have risen precipitously, which is further complicated by the fact that fewer drugs are commercially successful "Despite all this, we continue to press very hard because we recognize that there is unmet need and there are financial rewards for real innovation that can really help people, even if it's the exception rather than the rule," said Neil.

In recent years, public confidence in the pharmaceutical industry and in the regulatory system has eroded. This may have the effect, if the regulatory process is lengthened, of delaying the introduction of innovative products or adding additional expense to the process and ultimately the final approved product or medicine. "We can't sacrifice rigor, and no one is suggesting that, but we need to recognize the consequences of raising the regulatory bar."

To improve both productivity and regulatory certainty, said Neil, work needs to continue on understanding basic biology. "It's not easy," he said. "This is going to require a lot of collaboration between industry and academia." In addition, a new tool set is needed for drug discovery and development as it relates to translational medicine, and these tools need to be customized for particular diseases to increase the likelihood of an efficacious therapeutic agent for a particular disease.

The United States should invest in an infrastructure akin to the Internet or the interstate highway system in which it would be possible to enroll patients in clinical trials much more rapidly, whether for drug trials, observational studies, investigations of medical devices, or other research. Only 3 percent of cancer patients enroll in clinical trials today. "We make it inconvenient for them. Do we need an institutional review board in every university? Why can't we have national review boards? Why can't we have national safety monitoring committees? Why can't we bring the cost down and make the efficiency much better? Why can't we include patients of color, women, and older people? We're not getting those people today."

The nation needs a more sophisticated and effective safety and performance monitoring system for drugs once they enter the market. And, most important, said Neil, health care providers need a system to provide them with the latest information at the point of care to help them make the best possible decisions for each individual patient.

FDA regulates 25 percent of the U.S. economy, representing over $1 trillion worth of spending and a third of all the imports, with just 11,000 people and a $3 billion budget. "They need help," said Neil, including contemporary tools and techniques for pre- and post-marketing evaluation. They also need new risk assessment tools and much better engagement of patient communities.

"The standard way of looking at this is to talk about risk and benefit," said Neil. "What is the benefit of the treatment? What is the risk? . . . I think a better way of looking at this is risk and risk. There is a risk of not treating a disease. What is that risk? Then there is a risk of treating the disease. What is that risk, and what does that risk ratio mean in the minds of the patient?"

RESEARCH AND OUTCOMES CASE STUDY: PEDIATRIC HIV

Laura Guay, Research Professor at the George Washington University School of Public Health and Health Services, provided a perspective on research funding and evaluation by a philanthropic foundation. As vice president for research, she spoke about the work of the Elizabeth Glaser Pediatric AIDS Foundation, founded in 1988 to prevent HIV infection and eliminate AIDS among children in the United States and abroad through research, advocacy, and treatment programs.

Early on, the foundation studied how children are infected, how many children are infected, and why children are infected, chiefly through "scientist awards" to encourage young investigators to develop their careers in this less known field. The awards have provided $750,000 to individual investigators for capacity building rather than for specific research questions. Since 2007 the foundation also has made operations research grants to improve treatment program design and scale-up. As is often the case with medical research, there are obstacles to the delivery of science into the field, especially in developing countries, Guay noted.

Guay said that the foundation chooses innovative studies that are less likely to be funded through NIH. "Why isn't this fundable by the NIH" is one question on its application. For example, while funding for HIV vaccine-related studies is plentiful, very few of these funds focus on a vaccine in infants born to breast-feeding mothers. The foundation also may fund young investigators who do not have sufficient credentials to compete successfully for NIH grants.

To measure the impact of its research investments, the foundation needs performance metrics for deciding the impact of that funding, Guay observed. For example, an important question has been how awards have leveraged additional funds. Dating from the first funding of scientists in 1996, the foundation identified early leading scientists, which has generated an "exponential increase" of originally small investments over time.

Guay described two examples of the foundation's operations research, both influential in improving maternal HIV diagnosis and antiretroviral treatment in African countries. The first involved a controlled experiment in training nurses in the appropriate follow-up to a positive diagnosis of the infection. The second involved an experiment in rapid syphilis testing in connection with rapid HIV testing. Both projects provided evidence for methods of identifying more infected women, preventing transmission to their babies, as well as attracting men for testing and treatment.

Guay concluded her remarks with an illustration of successful application of research results that was recalled several times later in the workshop (Figure 4-1). In 1994, when research results showed that treatment of pregnant women with antiretroviral drugs could prevent babies from being born infected with HIV, the number of perinatally acquired AIDS cases dropped from approximately 900 per year in the United States to virtually zero over the next decade and a half. This dramatic outcome depended on progress dating to well before that date in human capacity, laboratory capacity, and clinical capacity, Guay observed. It is important to consider "all of the pieces that had to be in place" as "we continue to eliminate pediatric HIV in the rest of the world."

Sampat asked how the Glaser Foundation allocates funds among basic studies, vaccine development, and operations research, and Guay said that in the early years the foundation considered its funds unrestricted. But as more work has been funded by NIH and others, the trend has been to "donor-driven" funding for particular projects or areas. Because "people believe NIH has a lot of money," it is harder to raise foundation funds for basic research. And the biggest challenge, said Guay, is "we have a lot of science we haven't figured out how to deliver."

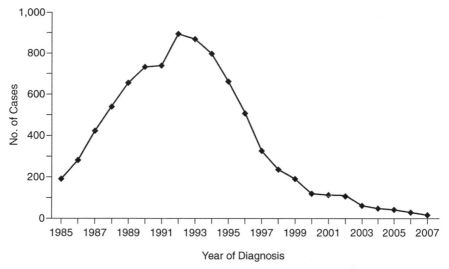

FIGURE 4-1 After it was shown that treatment of pregnant women with antiretroviral drugs could prevent babies from being born infected with HIV, the number of perinatally acquired AIDS cases in the United States and dependent areas dropped precipitously.
SOURCE: Guay, 2011

DISCUSSION

Given the "exceptional return" of HIV-AIDS research in the United States, "we need more case studies on failures" to figure out why some research avenues have not been more productive, said Sampat. Also, in evaluating the outcomes of public sector health research, it is hard to aggregate across disease areas. Sampat cited the Research, Condition, and Disease Categorization (RCDC) database started by NIH in 2009, which reports on 229 diseases and research areas of interest to Congress, but "these are not necessarily the diseases of historical interest to economists and policy analysts."

Kai Lee of the Packard Foundation, who spoke later in the workshop, asked if the data show "there is a lot more to be gained in the biomedical field from behavior-focused research?" He noted that Freeman, Citron, and Guay had all suggested the importance of human and institutional elements to outcomes. Freeman agreed that institutional and behavioral factors are important in the environmental area; they appear frequently on NIH's list of grand challenges as well. For example, the biggest success in preventing cancer has been behavioral, with regulatory, marketing, and other factors all working to reduce smoking.

Citron pointed out that that over time the optimal ratio of biomedical research to behavioral change could change. For example, though cigarette smoking has declined, many aspects of diet still need to change to improve health.

5

MIXED MARKET AND NON-MARKET IMPACTS OF RESEARCH

Like the benefits of research to health, many other research benefits may not be reflected or only partly reflected in market transactions but have enduring national importance. Examples include contributions to national defense, agricultural innovation, environmental protection, and the sustainability of natural resources. Economists have tools to measure the economic effects of non-market benefits, yet these tools may not always capture the full extent of those benefits.

Five speakers at the workshop examined these non-market impacts from very different perspectives, yet their observations had some intriguing commonalities. Foresight, leadership, and risk are all involved in pursing research with difficult-to-measure but very real benefits.

MEASURING PROGRESS TOWARD GOALS IN AGRICULTURAL PRODUCTIVITY

The Bill and Melinda Gates Foundation is a private foundation focused in part on improving health, reducing poverty, and improving food security in some of the world's poorest countries. It engages in what Prabhu Pingali, Deputy Director of Agricultural Development at the foundation, termed strategic philanthropy. The foundation establishes a set of clear goals; identifies the pathways, partners, and grants necessary to make progress toward those goals; and then measures progress toward those goals. In its Agricultural Development Program it focuses on doubling the productivity of farming by small landholders (less than two hectares) in sub-Saharan Africa and South Asia.

There is a rich history of metrics in agriculture development over the past several decades, Pingali observed. Since the Green Revolution, agriculture development specialists have been tracking the adoption and diffusion of modern varieties of the major table crops, so they know the extent to which modern wheat and rice crops have been adopted by farmers in developing world and the connection of that diffusion to productivity growth. This work also has shown that the rates of return to crop R and D in the developing world have been consistently high—on the order of 50 percent or more. Furthermore, these high rates of return have also high pay-offs for U.S. agriculture. For example, according to a study by Philip Pardey and colleagues for the International Food Policy Research Institute (IFPRI,1996), from an overall investment of $71 million since 1960 in wheat improvement research at the International Maize and Wheat Improvement Center of the Consultative Group on International Agricultural Research (CGIAR), the U.S. economy realized a return of at least $3.4 billion and up to $13.4 billion for the period 1970 to 1993. From a total investment of about $63 million since 1960 in rice research at the CGIAR's International Rice Research Institute, the United States gained at least $37 million and up to $1 billion in economic benefits from 1970 to 1993, according to the same study. "The bottom line," Pingali concluded, "is that international crop improvement research has had high pay-offs, not just for the countries where the work was targeted but also high pay-offs back to U.S. agriculture."

For small landholders in the developing world the chief crops are rice, wheat, maize, sorghum, millet, and cassava. For each crop the foundation has set clearly defined output targets that it expected grantees to achieve. For example, an output could be the release of a particular variety of maize that is tolerant to drought, or it could be the number of farmers in a given area who adopted a variety over a period of time. For grants across the entire food chain from seed to the consumer's plate, defining outputs becomes increasingly complex. Outputs for the use of fertilizer are straightforward, but what are outputs for fertilizer policies? Nevertheless, once specified by the foundation, grantees are expected to apply a set of indicators to track progress toward achieving those outputs.

The foundation has also sought to measure the extent to which its $1.7 billion agriculture investment over four years has reduced hunger and poverty. "Just adding up the outcomes from ways to monitor grant making does not necessarily get us to the answer," said Pingali. To address this problem, it has set up a randomly sampled household survey across Sub-Saharan Africa that is nationally representative and stratified by the agro-ecologies present in each country. It is now in the process of collecting detailed household data on production practices, technologies used, income, nutrition, and health and education status for about 25,000 households in seven countries in Africa and hoping to extend the survey to other countries. Visits to each household are occurring from one to two years apart over a 15- to 20-year period. "We can track changes that are taking place in African households over a long period of time and then track the contribution of productivity improvement to household welfare and the relationship between those two over this long period of time," said Pingali. "Of course we won't be able to attribute those changes specifically to our efforts, but I don't think that matters as long as we can show that there's progress toward achieving our ultimate goals of hunger and poverty reduction."

INVESTMENT DECISIONS AT DUPONT

As it enters its third century, the DuPont Company is undergoing a transformation that is bringing biology into a product mix based on traditional chemistry, said Richard Broglie, Director of Research Strategy at DuPont Agricultural Biotechnology. Its investment decisions are informed by four global megatrends: increasing food production; decreasing dependence on fossil fuels; protecting people, assets, and the environment; and growth in emerging markets. These trends derive in part from population projections. Global population is expected to exceed 9 billion by 2050. Feeding that number of people will require an increase in food productivity of 70 percent, Broglie observed. To meet this need, the majority of DuPont's R and D investments are aimed at adding new traits into crops to increase and protect yields, improving farm input efficiencies, and increasing the end use value of either the grains or the non-harvested crops.

DuPont measures the results of its investments in several ways, said Broglie. It tracks the number of new products introduced (with 1,786 new products produced in 2010), the revenue generated from those products, and the number of patents filed. The first two measures are more important than the third, said Broglie, since patents increase the probability of developing a product but do not necessarily give rise to products.

In the agricultural biotechnology area, a stage-gated approach for R and D decisions is used that progresses from discovery to proof of concept to early and advanced development to pre-launch to launch. This framework allows the company to balance its research investments across a diverse portfolio and over an extended period, since the development of a new crop trait can take 15 years or longer. It also helps balance investments against regulatory costs, which can be anywhere from $100 million to $150 million. At each stage, decisions involve people from the technical organization, the legal organization, the regulatory group, and the marketing group.

CHALLENGES IN QUANTIFYING RESEARCH VALUE IN AGRICULTURE

An economic cost-benefit analysis is an interesting problem but can be very difficult to implement, according to Michael Roberts, Assistant Professor of Agricultural and Resource Economics at North Carolina State University. In the case of research, economic analysis has shown that it is the main source of productivity growth. It is also a public good, which means that one person's use of research findings does not diminish its value to others and it is difficult for someone who has it to keep other people from using it. Because of these features, the private sector tends to do too little research, and there is a clear public role in funding research. However, to know how much to invest and how to set research priorities, the costs and benefits of different kinds of research must be weighed.

"This is a challenging conceptual problem," said Roberts. Research has many possible outcomes that economists might model as random. The range of potential outcomes is large, sometimes unintended, and probably unquantifiable. "We probably can't even imagine what the potential outcomes are of any individual research project." Many drugs used today are by-products of efforts to do something else, which reflects the uncertainty of research.

A Pest Forecast System as a Model

A recent research project in which Roberts was involved highlights some of these difficulties. In late 2004, a spore that causes soybean rust, which was then prevalent in South America and much of the rest of the world but not in the United States, landed on the shores of the Gulf Coast. The spore did not reduce yields much but it greatly increased costs because of the need to apply fungicides. The USDA coordinated its experiment stations to set up sentinel plots throughout the United States and monitor for soybean rust to track its spread. Also, an aerobiologist modeled how the spores move around on the winds, with a website reporting the overall results. Farmers could use this information to decide whether to spray fungicide on their soybeans or not.

The USDA's Economic Research Service sought to determine the value of this research. It took into account three key components: (1) prior beliefs about the amount of risk, (2) the amount of preventable losses, and (3) how well the information system resolves uncertainty. With no information, farmers will sometimes spray when unnecessary or not spray when needed. With perfect information, farmers will always make the right decisions. In the real world, partial information is available. For example, farmers had the option of carefully monitoring their fields, spraying the preventive fungicide, or monitoring their fields and spraying a less effective and less costly fungicide.

This range of scenarios made it possible to model the value of information, in terms of dollars per acre, against the range of prior beliefs about the possibility of infection. The model exhibited peaks of value that represented particular probabilities of beliefs about infection where a rational farmer would switch from doing nothing to monitoring and then to applying the curative fungicide. "You get these peaks right at the decision points because that's where you're most unsure about what the right decision is to make, and a little bit of information goes a long way at those points."

The USDA researchers concluded that the model had value. However, it was still crude. The model depended on an extraordinary simplification of reality and key simplifying assumptions. It had the potential to resolve subjective uncertainties, yet the quantifiable benefits were still difficult to determine and sensitive to the assumptions made.

In light of these limitations, Roberts was pessimistic about valuing individual research projects. However, other strategies may be more productive. For example, it may be possible to value research programs rather than projects. It also may be possible to value canonical examples, such as the development of hybrid corn, which depended on the work of a few key researchers. Finally, it may be possible to value projects and projects in retrospect and adjust research priorities accordingly.

Climate Change Projections

Roberts has been doing research on the effects of climate change on the global crop system. A key finding has been that extreme heat is by far the single most predictive variable for crop yields. This finding could be used to build an early warning indicator that would allow societies to avoid some of the adverse effects of climate change, he said.

However, immense uncertainty continues to make the value of this research difficult to quantify. Research seeks to find low-probability events that have extremely high payoffs. Economists would say that the value distribution has a fat tail. In a totally different context, climate change could have a fat tail if it has a small probability of producing truly catastrophic events. Cost-benefit analyses for research need to be pursued, but in cases like these they may not be feasible, Roberts concluded.

MEASURING SUCCESS IN CONSERVATION

The three major questions raised by Irwin Feller at the beginning of the workshop are somewhat different in the context of a private foundation's decisions, said Kai Lee, Program Officer with the Conservation and Science Program at the David and Lucile Packard Foundation. The first question becomes how much a foundation should spend on science, which is a question that is ultimately answered by the trustees within the constraints of a foundation's mission and resources. The second question becomes how to allocate funding given the mission of the foundation. And the third question becomes which research performers should receive the funds from a foundation. In the case of the Packard Foundation, said Lee, program officers are looking for a very specific population of research performers— people willing to work with the foundation to contribute to informing the near-term decision making of entities, including public agencies, that will support the foundation's conservation mission. "That turns out to be a lot harder than you might think," he said.

The Packard Foundation made $236 million in grants in 2010 in four areas: population and reproductive health; children, families, and communities; local programs; and conservation and science, with the last of these categories accounting for $154 million in grants in 2010. For example, it supports the Monterey Bay Aquarium Research Institute, which is a major oceanographic institution created by David Packard in which scientists and engineers work together. It has a fellowship program in science and engineering for early career scientists. And it has other programs focused on oceans science, which is a major emphasis for the foundation. Although the amounts of research support it provides are small compared with federal funding for research, the foundation is a significant funder in the field of marine conservation.

In general, knowledge of oceans conservation is held by government agency staff members, academic scientists, and a growing cadre of scientists who work for non-governmental organizations that have varying degrees of advocacy as part of their mission. This knowledge has come to be a countervailing source of information for decision makers in the face of advocacy by resource users and developers, who also depend heavily on publicly funded knowledge.

The foundation seeks to link knowledge with action. While advancing conservation strategies, it also works to improve the use of knowledge in decision making. "In effect, what I'm trying to do is to foster a kind of 'learning by doing' by making grants and working with users and researchers," said Lee. Using this approach, real-time evaluation of outcomes is an essential component.

In the conservation field, the use of knowledge to inform action can be done in two possible ways. One is to bring knowledge to bear to support advocacy to achieve specific conservation ends. The problem with this approach is that knowledge becomes entangled in polarization. "There is a grave risk of damage to the credibility and legitimacy of science when it becomes entangled in that polarization," said Lee. "Nonetheless, science in support of advocacy has sometimes proved to be necessary and successful."

The second approach is not to support advocacy but rather to support decision making and learning. This tends to work best in a collaborative setting. In such a setting, science is part of a governance process to solve problems rather than part of a polarized process to try to change the rules. This use of science tends to reinforce existing institutions, but it also requires some conflict so that problems can be recognized and information being brought to bear by science can affect decisions.

Lee discussed the concept of adaptive management, which he described as the idea that the implementation of a policy should be understood as an experimental test of the hypothesis embodied in that policy. Such an experiment requires systematic monitoring of outcomes to determine the consequences, including unanticipated consequences, of a policy. "You want to do integrative assessment of that knowledge to build knowledge of the system that you're innovating in, the ecosystem if you like, to inform model building, to structure a debate, and from that to enable strong inference."

The science Lee seeks to support links communities of scientists with decision makers, stakeholders, residents, and citizens of an area who are used to making decisions without any information from science. It can be difficult to make this connection work, Lee observed, so often the foundation has tried to foster the emergence of boundary-spanning organizations. The foundation does this by emphasizing output-oriented grant making, in which it focuses on decisions makers at the outset. "We put a lot of effort into aligning users and researchers, and this is where the art of the grant maker gets called upon." The foundation presents prospective grantees with a set of questions to think about as they prepare their proposals. "We want them to understand and explain to us whether the situation is the right one. That is, is there an opening for new knowledge to actually cause changes in action." The process can be burdensome, with the foundation identifying specific indicators and closely monitoring their progress. "The objective is to allow us to learn about the types of short- and medium-term interventions in which the foundation can have the greatest impact."

NATIONAL SECURITY BENEFITS

Richard Van Atta, Senior Research Analyst at the Science and Technology Policy Institute, pointed out that national security is also a societal value with a very fat tail. The value of national security can be viewed as infinite, or at least as binary, in that the United States has it or it does not.

Similarly, defense research can have immense payoffs that are difficult or impossible to predict. For example, a relatively modest investment in gallium arsenide monolithic microwave integrated circuits for signal processing led to the development of a technology that is now used in every cell phone around the world.

Despite these uncertainties, the Department of Defense still has to assess the effects of research investments on national security as a way of making decisions. Research in the Department of Defense is purpose-driven, Van Atta said. The nation relies on the technological superiority of its armed forces to maintain its position of world leadership. The question then becomes: How can the value of technological superiority be assessed in terms of desired outcomes? "You can't defend everything against everybody, so you have to make choices."

The Department of Defense conducts this assessment by establishing a national security strategy and then relating technologies to the strategy. In doing so, it differentiates technologies according to different objectives. *Core technologies* refer to longstanding traditional capabilities, such as explosives and propulsion. *Critical technologies* refer to revolutionary or transformational technological changes. *Emerging technologies* occupy the forefront of knowledge and have the potential to be critically important but have not yet been fully developed. *Process and manufacturing production technologies*, such as process controls for nanotechnology, underlie other developing technologies. *Enabling or cross-cutting technologies* are capabilities that everyone wants but does not want to pay for. In this case, different organizations may be devoting insufficient effort to the technologies, and these efforts need to be scaled up to produce a technology that will have a substantial impact.

In all of these cases, technologies need to be managed in increasingly difficult and complex technology environments. This management requires the establishment of goals and purposes. For example, NASA has an approach called GOTChA, for Goals, Objectives, Technology Challenges, and Actions or Activities. Under this approach, activities are organized toward goals by focusing on the questions "Are we getting there?" "Are we there yet?" "How far have we gotten?" "Do we put more in or don't we?"

The DARPA Approach

DARPA is the best known organization within the Department of Defense for developing high-payoff high-risk technologies, observed Van Atta. When George Heilmeier became Director of DARPA, he imposed what came to be known as the Heilmeier Criteria. These were basically a set of management questions that asked: What is the purpose of doing this research? What difference will it make if it succeeds? How would you know if you are succeeding? What are your midterm criteria for assessing it? And what are your milestones? When researchers responded to these questions by saying, "We're scientists; we can't tell you those answers in advance," Heilmeier responded, "You will if you want my money."

This approach to assessment is oriented toward research designed to meet specific identified needs, said Van Atta. That begs the question of how to define these needs and how to link them to requirements that have not yet been specified. "We know what the requirements are for today," said Van Atta. "What are the requirements for five or ten years from now in the security world?" During the Cold War, the requirements changed slowly. "Today the security environment changes faster than we can develop our S and T plans. It's more like the business environment," which requires that technology development be managed in a different way than in the past.

PUBLIC PROBLEM SOLVING

Public Agenda is an organization devoted to bridging the gaps between leaders and the public and also experts and the public, said the organization's president, Will Friedman. By measuring and then working to reduce these gaps, Public Agenda and similar organizations engage stakeholders and help people come to terms with issues.

Public Agenda does considerable public opinion research to find out how people are looking at problems. It also conducts public and stakeholder engagement and communications to set in motion collaborative processes. It has worked on many issues, including energy, the environment, and health care.

The organization tends to become involved in complex societal issues that involve both science and politics. In these cases, people need to make value judgments and adapt to change. Public participation may not be needed to enact a policy, but the lack of participation can lead to backlashes that undermine a policy. Consequently, the challenge for Public Agenda is usually how to create the conditions that allow the public to come to terms with complex, science-intensive issues.

The way the public wrestles with issues and comes to hold certain positions is different than how experts wrestle with issues, Friedman said. The public learning curve involves three stages, beginning with a consciousness raising period. For the public to come to terms with an issue, they need to develop a sense of awareness and urgency about that issue. The public then engages in a process of working through an issue. Many barriers can impede this process, including a lack of urgency, wishful thinking, misperceptions and knowledge gaps, and mistrust. Overcoming these barriers requires strategic facts, appropriate choices, and time. "The real art and science here is to be much more precise, not in terms of your desire to manipulate the public to have the opinion you want them to have, but rather to help them figure out where they actually stand and what's important to them."

In the case of climate change, for example, surveys have shown that the public has become *less* likely over time to view climate change as serious. Further work showed that people were not getting the message that scientists thought they were delivering. The public tends to frame the issues in terms of bread and butter issues— for example, that gas prices and reliance on imported oil are more serious threats than climate change. They have a great deal of wishful thinking, and the issue has become polarized by politics.

Science has a role in helping the public grapple with such issues, but it may not be the role many scientists assume. Their most common mistake it to demand that the public become junior scientists. As a result, they overload people with technical detail without considering what information the public is ready to receive at a given time. "Science literacy is well-intended and education is a good thing, but it does not necessarily help people grapple effectively with specific issues at specific points in time," said Friedman.

Science's most important contributions are to lead the charge on the technical side of problem solving while informing public deliberation in critical ways. Science can help clarify the choices the nation needs to make. It can help people understand the implications of different solutions and the tradeoffs involved. Public Agenda uses a tool it calls a choice framework that presents people with a few strategic bits of background information - "not too much, but just based on research about what it is that people need to begin to get into the issue." It also studies the framing of issues in different ways to help people deliberate more effectively. The choice framework "can help people learn quickly and shift from a non-productive, circular reasoning and non-exploratory dialogue to one where they are working off each other, thinking about solutions, and generating really interesting questions."

DISCUSSION

During the discussion period, Van Atta was asked how to build institutional support for entities such as DARPA that are institutionally disruptive. The best approach, he said, is through top-down leadership. For example, the impetus for stealth technologies came from the Secretary of Defense and depended on his vision and strategy in pursuing a new technology. "If you're going to do something different, you've got to do something different."

Broglie was asked whether DuPont has a strategy for releasing research results into the public sphere when they do not lead to marketable products but could nevertheless lead to important advances. The question is difficult to answer, he said, because there are many reasons why something might not progress through the commercialization pipeline. However, DuPont has worked with the Gates Foundation on crops for which it does not sell seed to improve the nutritional quality of grains. In other cases, technical dead-ends are publicly released to make information available that has public value.

Roberts was asked about liability considerations if a model leads farmers to make a decision that turns out to be mistaken or harmful. He agreed that for a model to be useful as a decision tool, it would need lots of supporting data. Also, through use the model would be refined.

6

IMPACTS OF RESEARCH ON THE LABOR MARKET AND CAREER DEVELOPMENT

Several speakers during the workshop contended that the most important influence of research is the training it provides for undergraduates, graduate students, and postdoctoral fellows, who then bring those experiences and skills into the workplace. Three speakers at the workshop looked specifically at that assertion. Economic analyses can reveal the value of these workers to the economy, while survey results can uncover the preferences and goals of workers and employers. However, many questions still surround the processes through which supply and demand interact.

R AND D SPENDING AND THE R AND D WORKFORCE

In the short term, the relationship between R and D and the workforce is relatively weak, said Anthony Carnevale, Director of the Georgetown University Center on Education and the Workforce. But in the longer term, the relationship can be much stronger.

Explaining the Residual

Economists explain economic growth and productivity increases in part by citing the development of human capital and investments in physical infrastructure. But those two factors explain only part of the growth of the economy. The residual— "between 65 and 40 percent, depending on who you read," Carnevale said— comes from advances in knowledge.

Many economists think of these advances in knowledge as being embodied in technologies, but in fact the residual consists of everything

49

that cannot be measured as a direct investment in the economy. Carnevale said that he preferred to think of advances in knowledge as the way people combine and use resources, whether human, technological, or otherwise. So advances in knowledge include the development of Walmart as opposed to mom and pop hardware stores, not just the direct effects of technology.

R and D Spending and Economic Growth

Connecting federal spending on R and D to these advances in knowledge is a difficult problem. For example, R and D directly involves a fairly limited number of people. About 1.4 million U.S. workers spend at least 10 percent of their time doing R and D, out of a total workforce of about 150 million people. (The former number includes social scientists, although the Center on Education and the Workforce typically does not include social scientists among workers in science, technology, engineering, and mathematics, or STEM.) The relatively small size of the STEM workforce explains why federal investments in research have relatively small short-term impacts on employment.

The STEM workforce engages in both research, which Carnevale identified as scientific investigations— and development — or the application of scientific knowledge. While research has sometimes led directly to technologies that are economically important, development is a much more important source of innovations, according to Carnevale. "Historically, science owes a whole lot more to the invention of the steam engine than the steam engine ever owed to science. That is, most of the development of economies occurs in application, not in labs." A strong argument also can be made, he said, that the economic value from development has been growing more rapidly than the economic development from research. "A lot of wealth creation in the world now has to do with process improvements, not so much invention." Even in industries such as pharmaceuticals, where discoveries lead to new products, the commercialization and distribution networks bring in much of the new revenue.

The Growth of the STEM Workforce

The STEM workforce, which is larger than the number of people doing R and D, is growing, said Carnevale. Today, people who work in science, technology, engineering, or mathematics— not counting social scientists—represent 5 percent of the workforce, and this percentage is increasing.

The STEM workforce represents the endpoint of a long process of attrition, Carnevale pointed out. Many people with high mathematics scores in grade school and high school do not want to be STEM workers and do not pursue those subjects when they go to college. Among those who enter college declaring an interest in STEM subjects, many switch to other majors before they graduate. Even among STEM majors, many go into other careers. And among those who begin in STEM careers, many move out of the STEM workforce, especially after the age of 35.

In part, this attrition results from opportunities in other fields. Wages for STEM workers are relatively high, but the wages in other fields associated with high test scores in areas such as mathematics are even higher. Competencies developed in STEM fields are in demand in a large and growing share of occupations that pay well, which translates into many opportunities for people who have those competencies.

Also, workers who switch out of STEM fields tend to have values and interests that are different than those associated with STEM occupations, Carnevale said. Among STEM workers, the values and interests recorded by industrial psychologists are relatively narrow, whereas the values and interests in the general workforce are relatively broad, especially for high-achieving students who have many choices.

Given these observations, said Carnevale, the United States is going to have to rely more and more on foreign-born STEM workers. International diversity is now greater than the domestic diversity in the STEM workforce, and a healthy and productive STEM workforce will require focusing on both sources of diversity.

SURVEYS OF GRADUATE STUDENTS AND POSTDOCTORAL FELLOWS

Existing surveys reveal valuable information about the career trajectories of graduate students, postdoctoral fellows, and early career scientists and engineers, but they also have many limitations. Henry Sauermann, Assistant Professor of Strategic Management at the Georgia Institute of Technology, profiled existing surveys and described a new survey that he and a colleague conducted that has provided valuable additional information.

Existing Sources of Data

Several different data sources provide information on the aggregate flows and stock of scientists and engineers. The National Science Foundation's Survey of Earned Doctorates, which Ph.D. recipients fill out when they graduate, provides much valuable data and now includes financial information such as salaries, at least for the people who have job offers. In addition, the Survey of Doctorate Recipients (SDR), the National Survey of Recent College Graduates (NSRCG), and the National Survey of College Graduates (NSCG) – NSF's other personnel surveys— all provide important data on the stock of intellectual capital available to the economy. In addition, some information on postdoctoral fellows is available through the Sigma Xi survey and through the SDR.

Once students become active scientists, they begin to produce publications and patents, which can be used to track where people go, what they do, and the extent of their collaborations. Finally, a new federal data collection program, STAR Metrics (discussed in detail in Chapter 8) collects information on funding for public research and the extent to which that funding is used to support postdocs, Ph.D.'s, or other students.

Sauermann described what he called his "wish list" of data that would be very useful to have. For example, when a student reports moving from Stanford University to a company, the move reflects a labor market transaction. But the data do not reveal what the student or the company wants. More information is needed on both sides to know how well the job market is operating. On the supply side, the data might include aspirations, intentions, and skill sets. On the demand side, what kinds of jobs are open and what kinds of skills do firms need? For example, an ongoing argument, said Sauermann, is over whether the United States has too few scientists who know something about business and who can work in larger teams and companies. "It's a question about the match between the training that individuals receive and what is required on the demand side."

It is also important to understand more about how the labor market works, Sauermann observed. Supply and demand might match in the aggregate, but there may be great inefficiency in that process. Not every job seeker knows all the potential employers, and not all the potential employers know about all the people they might hire. How do students collect information? Who tells them about different careers? To what extent do advisors know what an industry or government job entails? All of these questions are important.

It also would be interesting in know more about the training experience itself and how training translates into future career outcomes, Sauermann said. An ideal data set would track individuals when they enter a Ph.D. program, ask them why they are seeking a doctorate, track their learning experiences, and determine how their experiences changed their intentions. "This is really important if you think of graduate school as the place that trains people and socializes people into becoming scientists."

Current data reveal very little about people who do not graduate. Do they consider their time in graduate school to have been wasted? Was it good for them to realize that graduate school might not have been a good fit? How do institutions make selection decisions?

Finally, current data provide little information on people who earn doctoral degrees outside the United States, though some efforts are under way to get more data about these individuals.

A Science and Engineering Ph.D. and Postdoctoral Fellow Survey

To learn more about the attitudes and actions of graduate students and postdocs, Roach and Sauermann (2010) conducted the Science and Engineering Ph.D. and Postdoc Survey (SEPPS) at 39 leading research universities in the United States. They collected contact information for 30,000 individuals, conducted the survey in the spring of 2010, and had about a 30 percent response rate. The survey focused on advanced Ph.D. students who had passed any necessary exams and postdocs in the life sciences, chemistry, physics, engineering, and computer science.

One question they asked was, "Thinking back to when you began your Ph.D. program, how important were the following factors in your decision to pursue a Ph.D.?" Respondents agreed more strongly with the statements that they were always interested in research, were curious to learn about a specific field, or needed a Ph.D. for a desired career. They agreed less strongly with the statement that they admired the status of people holding Ph.D.'s, and they agreed least with the statement that they had difficulty finding another job. Research "is a career that people consciously choose as opposed to being forced into it because there's nothing else to do," Sauermann concluded. Also, although some foreign graduate students and postdocs agreed that getting a Ph.D. offers opportunities to secure a visa, on average this motivation did not rank highly.

When postdocs were asked the same question about their fellowships, they agreed most strongly with the statements that a postdoc would increase their chance to get a desired job and deepen their skills in a particular area. They agreed moderately with the idea that a postdoc gave them more time before deciding on a career and agreed less strongly with the statement that they had difficulty finding another job.

When asked about their current funding sources, between 70 and 80 percent responded that they were funded by federal sources. About 60 percent got university fellowship and assistantship funding. Private foundations were quite active, especially in some of the fields, while very few respondents received industry funding. Postdocs in the biological and life sciences got fewer university fellowships and assistantships but more industry funding.

When postdocs were asked, "How involved were you in securing your most important source of funding?" respondents in the biological sciences averaged 50 points on a scale from 0 to 100, while people from physics averaged 38, people from computer science 29, people from chemistry 38, and people from engineering 39.

The survey asked whether their research contributes fundamental insights or theories, or whether it creates knowledge to solve practical problems, with people being allowed to respond affirmatively to both questions. They were also asked whether they were interested in doing basic research or applied research later in their careers. Among the life scientists, people who got federal funding were much more likely to be engaged in basic research than people who did not get federal funding. Similarly, those getting industry funding were much less likely to be engaged in basic research than those who did not. People receiving funding from foundations were also more likely to be engaged in applied research.

Interestingly, there was not much relationship between funding source and career aspiration or what people wanted to do later. The only exception is that people who got industry funding tended not to be interested in working in basic research later.

Two other question asked, "How much freedom do you have in choosing your research topics?" and "How much freedom do you actually have in influencing the direction of your research projects?" People with multiple funding sources reported an increased level of choice in terms of what they wanted to work on as well as in terms of deciding how they want to work on these things. The only individual funding source that made a big difference was foundation funding, where people felt much more freedom in their choice of research topics. "Presumably, that's not because the funding makes them free, but because they have a pet project, or they're enthusiastic about something, and they go apply to different foundations. . . . In that sense, foundations seem to provide a lot of freedom— not because people get their money first and then choose but because they choose first." In contrast, industry funding tends to have a slightly negative impact on freedom, but only for postdocs.

Finally, the survey asked about the types of jobs respondents found most appealing, whether teaching at a college or university or doing research at a college or university, a government research institution, an established firm, or a startup (Figure 6-1). Most of the respondents in the life sciences wanted to have a faculty R and D job, with 50 percent finding that the most interesting career. Physicists and computer scientists rated that option even higher, but chemists and engineers had less interest in a faculty R and D position and more interest in R and D jobs at established firms. People who received industry funding were less interested in a faculty research career and more interested in working either for a start-up or for an established firm.

The experiences people have during their education shape their involvement in the labor market, Sauermann concluded. "We need to understand more of what these labor market processes look like to see how we can direct or change, if we want to, these labor market outcomes."

THE COMPLEX NETWORK OF SKILLS AND INVESTMENTS

Recent discussions of U.S. science and technology policy have emphasized the concept of global competitiveness. As James Evans, Assistant Professor of Sociology at the University of Chicago, pointed out, this concept inevitably poses the question: What is a globally competitive STEM workforce, and how does the government best invest in developing this kind of workforce?

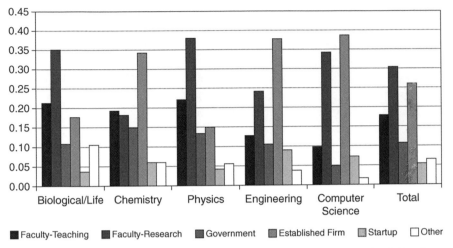

FIGURE 6-1 When postdoctoral fellows were asked "Please rank the following careers from most likely to pursue to least likely to pursue," Ph.D.'s in the biological and life sciences, physics, and computer science were more likely to favor faculty teaching jobs, while chemistry and engineering students were more likely to opt for jobs with established firms.
SOURCE: Sauerman, 2010

Competitiveness as Size

One framing emphasizes the much repeated concerns about the supply or size of the STEM workforce. For example, in a 2007 op-ed article in the *Washington Post*, Bill Gates wrote, "Demand for specialized technical skills has long exceeded the supply of native-born workers with advanced degrees, and scientists and engineers from other countries fill this gap. This issue has reached a crisis point." This framing produces a one-dimensional indicator of competitiveness that is fairly easy to measure, said Evans. However, with only 5 percent of the world population, the United States inevitably will drop below the 35 to 45 percent of global science and engineering activity that it retained through the end of the twentieth century. As the world continues to develop, more countries will be producing more scientific activity, and these scientists will receive more publications, more citations, and more attention.

Existing measurements of the STEM workforce are closely cued to size, Evans observed. Inputs to the workforce include the gross amounts spent on training grants and an unknown proportion of research grants spent on personnel in training. Outputs in surveys such as the SED, SDR, and STAR Metrics are the numbers of doctorates, the sectors of their jobs, their incomes, and self-reports of activities and outcomes (such as articles and patents). Given these measures, it is impossible to assess the efficiency with which the system matches inputs with outputs.

Competitiveness as Efficiency

Another framing is to think of competitiveness in the STEM workforce as efficiency in producing a sufficient supply of the skills in demand. From this perspective, the United States can be seen as the most efficient investor in science and engineering skill. Wages for STEM workers have been largely flat, said Evans. Reports of low supplies of scientists and engineers typically come from hot industries and from potentially self-interested parties, suggesting that there is no undersupply of skill. In fact, there may be an oversupply of skill or an oversupply of the wrong types of skill.

This framing leads to a more nuanced concern about the efficiency or the relevance of training investments in the STEM workforce. From this perspective, the relevant inputs are the size of the training investments and the relevant outcomes are the incomes of STEM workers, assuming that the market is clearing. But to make such an assessment, improved measurements would be needed. The first such improvement would be the educational components of research grants. The second would be improved information about STEM workers, such as some of the information described in the previous presentation. Measurements of efficiency also would require a better sense of preferences to judge the elasticity of individual human capital investments. For example, how much is it worth for students to have control over the subject of their research? Some natural experiments have yielded information on this issue. For example, when the size of a research grant goes up, the student response goes up in an approximately linear fashion. But real experiments should be organized, Evans said, because the presence of confounders can make natural experiments hard to interpret.

The problem with this framing is that it typically responds to past rather than future labor needs, Evans noted. For example, this perspective has motivated initiatives such as the Alfred P. Sloan Foundation's advocacy of programs that award the professional science master's as a terminal degree. But this effort may undervalue the doctorate, even if society or U.S. companies benefit more from a doctorate than does the recipient of that degree.

Competitiveness as Quality

A third framing equates competitiveness with quality. From this perspective, the United States can be seen as the elite global supplier of science and engineering skill, Evans observed. This indicator of competitiveness is very difficult to measure because it has such a high dimension. It also renders obsolete the idea of thinking about competitiveness in terms of a labor market. Instead, actual skills and their actual and potential value must be considered within the broader system of innovation. Researchers and their contributions can no longer be treated as independently and identically distributed. Even bibliometric methods are inadequate, because particular articles and patents fit within the system in certain places, and understanding those places is the key to the allocation problem. "When we open the box of content, instead of just measuring the numbers of papers, we have to look at the papers, we have to look at the content, and it's a daunting exercise."

Coauthorship and citation networks are one way to measure the contributions of individuals, though "it's not clear how much insight" they can produce, said Evans. Authors and papers can be identified as more central or more peripheral. Visualization techniques also make it possible to determine how clusters are linked together to form modules in a network. In addition, natural language processing and machine learning can increasingly discern the landscape in millions of papers to identify features of those landscapes. Together, these techniques "can give us a much richer and more powerful view of the value of investments," said Evans.

Doctoral STEM Education

Students who undergo a doctoral education emerge with a specialized set of skills and techniques, including meta-techniques, such as being able to design a research project. This observation raises several linked questions: What is the role of deep, specialized knowledge in exploring new knowledge or skills? What is the role of social networks developed or entered into through education in spreading knowledge or skills? And what is the role of interdisciplinary laboratories in managing novel combinations of knowledge or skills?

Evans studied these questions through an investigation of almost 20,000 publications involving *Arabidopsis thaliana* (a small flowering plant used as a model organism) in which he identified principal investigators, organizations, subfields, countries, genes, gene products, methods, and metabolites used. He found that the more persistent researchers were within these identified terms, the more central they were within the coauthorship network. At the same time, with these researchers it was more likely that industry collaboration and funding would influence their work to become more theoretically unexpected. In essence, government sponsorship encouraged validation and moved work toward the center of the network. Industry sponsorship encouraged novelty and pushed work toward the periphery of the network.

"This suggests an interesting and important complementarity between government and industrial efforts," Evans concluded. Governments sponsor hubs of knowledge, while industry involvement encourages the exploration of high-value novel combinations.

Network analysis of geographic localization also has shown that knowledge flows within communities and within firms. Furthermore, many ties in the biosciences are formed through doctoral committees and communities.

The important point, concluded Evans, is that analysts need to look beyond labor markets to the relative values of skilled people. Investigating this issue will require linking individuals and their preferences with the papers and patents they produce. "Labor market issues cannot be separated from the content of science."

DISCUSSION

In response to a question from a workshop participant about the importance of the arts and humanities in generating economic value, Evans noted that he was very interested in the complex combinations of STEM knowledge and the arts and humanities in such areas as design. "It's silly to cordon those things off in the context, especially, of industry and productivity."

Sauermann added that many people do not work in the field in which they studied, and these numbers are especially low for the social sciences. "Many people are studying stuff they don't use. Maybe that's by choice. Maybe not. Again, I think it would be interesting to know."

A workshop participant asked about the tendency of professors to train students for positions in academia rather than industry, to which Sauermann replied that some faculty members are very active in industry and have their students work on industry grants. However, in a separate survey, he asked students what level of money, freedom, equipment, and so on they expected to have in different kinds of careers, and many more students marked "Don't know" when asked about start-ups and established firms than when asked about academia. "It could be that they don't search it out because they don't want to be in industry. [But] there is probably less information out there."

Carnevale added that the U.S. Department of Education is supporting the development of an online system that will collect information on all transcripts of students, including those in college and graduate school, and connect that information to wage records supplied by every employer in America. Currently, in 26 states, a student in a Ph.D. program in physics can find out how many of last year's graduates got a job, whether it was in physics, what their wages were, and the duration of their employment.

7

INTERNATIONAL PERSPECTIVES ON MEASURING RESEARCH IMPACTS

Measuring the impacts of research ranges from studying broad changes in public policy to tracking the influence of a certain research paper on subsequent publications in that field. Some of the newest techniques marry on-line data collection and databases with analytic tools, yielding a nuanced picture of research outcomes and the influence of funding dollars. At the workshop, speakers from the United Kingdom (UK), the European Union (EU), and Brazil shared some of their thoughts on recent evaluation methods and future goals. Measuring the effectiveness of research is a growing field precisely because of the scarcity of resources and the need for policy makers to demonstrate returns on investments around the world.

MEDICAL RESEARCH COUNCIL EVALUATION SYSTEM

The United Kingdom Medical Research Council (MRC) provides government funding for public, private, and university research in the United Kingdom. Science funding in the UK comes from the government, the private sector, and charities, and universities function on a dual support system, where money for staff and infrastructure comes from higher education funding councils and research councils designate funds on a project and program basis. Ian Viney, Head of Evaluation for the London-based MRC, outlined the council's efforts to measure and influence research impacts.

The MRC is focused on collecting comprehensive evidence regarding the progress, productivity, and quality of research output; supporting studies along the lines of those funded by the National Science Foundation's Science of Science and Innovation Policy Program

(SciSIP); encouraging researchers to maximize their "pathways to impact"; and adding the assessment of impact as a factor in allocating new funds to UK universities. In 2006 the MRC started using an online system called e-Val. The system, which replaces end-of-grant reporting, requires grant recipients to make online reports each year, resulting in structured feedback over the lifetime of a grant rather than a long report at the end summarizing years of progress. The evaluation is designed to track how scientists are influencing policy development and contributing to new products and interventions. In building the evaluation, the MRC asked questions intended to yield hard evidence of impacts, outcomes, and output, in addition to traditional tracking of papers and patents.

In two years of data gathering, more than 3,000 researchers have participated. The system has collected 70,000 reports representing feedback on £2 billion of MRC funding, or 92 percent of MRC expenditures in the last four years. In 2010 the evaluation provided details on 5,000 active collaborations. Since 2006, MRC researchers reported over 130 citations in policy documents, 360 new products and interventions in development, 200 published patents, and 37,500 publications.

The online evaluation system helps the MRC link research outputs with the social, economic, and academic impacts of research. For example, one study done by the Health Economics Research Group, the Office of Health Economics, and RAND Europe (2008) focused on the return on investment for research on cardiovascular disease and mental health. Combined with data from e-Val, the study built a strong quantitative argument for investment in medical research in time for the change of party control of government in 2009 and the subsequent review of all government spending.

Monitoring policy citations and the influence of scientists in policy helps track progress over time and demonstrates how research translates to clinical practice, said Viney. The evaluations also have given context to case studies, which the council often uses to illustrate to the government the benefits of MRC funding. But it is not easy to encourage researchers to think about the ultimate objectives of their work and how to maximize their impact. Viney pointed out that the medical community is somewhat more accustomed to this, while other disciplines are more resistant.

The Research Councils UK (RCUK), which is made up of seven UK research councils that together allocate £3 billion each year to research, is keen to maximize the economic, academic, and societal impacts of research, and the councils are including information on these impacts in all of their funding applications. They ask researchers not to predict what the impact will be but simply to consider enhancing the potential influence of their research. A peer review process, "Pathways to Impact," is also designed to further this goal.

The Higher Education Funding Councils in the UK, which allocate £2 billion to university research every year, have moved in a similar direction. Due partly to pressure encouraging the Higher Education Funding Councils to look more closely at impacts, they implemented the Research Excellence Framework, which assesses research outputs, impacts, and the research environment at each university. The framework splits disciplines into units of assessment, defined as substantive bodies of research in coherent discipline groups. There are roughly 30 units of assessment. A pilot using expert panels to assess impact at 29 universities, with each university submitting case studies for two units of assessment, was considered quite successful. The panels found ways to assess the validity and significance of impacts across diverse disciplines, including clinical medicine, physics, earth systems, social work, and English literature. The panels will contribute 20 percent to the overall REF assessment, with the goal of increasing that contribution after 2014.

In the government's 2010 Comprehensive Spending Review, the MRC's evaluation helped protect the medical research budget in real terms until 2014 while the overall science budget received no inflation increase. For the MRC, this is a tangible example of evaluation influencing policy, Viney said. Other funding agencies are now looking at ways to imitate e-Val, and discussions are under way to harmonize and rationalize the data collection process with a view of generating a more UK-wide view of research output. Plans are also under way to commission more work on estimates of spillover benefits in the UK, rather than borrowing from U.S. estimates. Viney concluded that the government is focused on economic growth and leveraging investment, and the importance of describing, understanding, and assessing impact is becoming more widely accepted in the UK.

MEASURING IMPACTS OF RESEARCH FUNDING IN THE EUROPEAN UNION

When evaluating research, it is important to compare old and new approaches. Brian Sloan, Directorate-General for Research and

Innovation for the European Commission, discussed various forms of evaluation in the context of the European Union Framework Program, which supports European science, technology, and competitiveness. The program is designed to complement national programs, focusing on areas where national funding may not reach, and to encourage cooperation and coordination between countries. The program allocates funding for transnational research projects, and also for mobility so that researchers are able to travel from one country to another.

The current Framework Program, the seventh since 1984, has a budget of 50 billion Euros, or approximately $70 billion, which is 7 to 8 percent of European R and D funding. There are four components: *Cooperation, Ideas, People,* and *Capacities.* The *Cooperation* piece funds transnational research consortia. *Ideas* funds national teams that compete across the European Union. *People* funds mobility. *Capacities* provides funding for infrastructure. Within each of these divisions is a range of different science and technology fields.

Traditional methods that the Framework Program has used to evaluate the impacts of research include interviews, surveys of program participants, and expert panels. But Sloan pointed out several challenges inherent in these methods. Surveys can be a burden to participants, especially when long and detailed answers are required. This can influence the quality of their response. Response bias and partial responses are also a concern. In addition, because most research projects have various funding sources, it can be difficult to attribute specific findings directly to EU funding. While these methods are still quite valuable, it is worthwhile to look at new approaches.

New methods include what is called linking and ex-ante modeling. Until recently, it was difficult to identify recipients of EU funding by linking into bibliometric databases, but in 2009 it became possible to search grant activity and funding acknowledgements in the Web of Science database and therefore accurately identify not only program participants but their affiliates. Using the database in this way allows for assessment of research output and comparison with other projects, national averages, and world averages. There is also a built-in control group, which is lacking in surveys or participant interviews. Using bibliometric data, it is possible to map co-publication or track which disciplines publish most within the various programs.

This type of evaluation is particularly relevant for the Framework Program, as one of its goals is to measure the results of funding against other transnational endeavors. It is also possible to measure the effects of distance or language on collaboration, and evaluate whether the program is succeeding at connecting people and regions that would not otherwise be brought together.

Another approach the program took was linking with the Community Innovation Survey, a harmonized questionnaire that surveys 40,000 firms across 30 European countries. The survey looks at innovative outputs and activity, R and D spending, patents, cooperation, and new products. Included in the survey were questions asking whether firms had received any EU funding from 2002 to 2004 and whether they had participated in the Framework Program. The responses provided crucial data that could then be used to compare Framework Program participants with other researchers, controlling for variables such as company size and sector, and discern whether the program increases collaboration and productivity.

The commission also found ex-ante evaluation to be a useful tool when applied to the Framework Program. The European Commission produces an ex-ante impact assessment report each time it develops new funding programs, explaining what problem is being addressed, why the government and in particular the EU must intervene, the objectives of the program, and what policy options have already been considered. For each option, the assessment also includes predictions of economic, social, and environmental impacts.

Using an econometric model, the commission used a similar approach to assess macroeconomic impacts of the seventh Framework Program up to 2030 under various scenarios. The model predicted effects of the program on exports, imports, research, GDP, employment, and a range of other indicators. Again, like bibliometric data, this approach allowed for comparisons and manipulation of data, as well as bringing up potentially interesting and important developments that may not otherwise have been recognized.

Ex-ante evaluation and linking provide another angle on measuring research outcomes and impacts. Because official statistical surveys provide such a large amount of reliable data, sophisticated analyses can be done of networking effects that cannot be captured from participant surveys. Sloan emphasized the potential of such approaches to yield further progress in the future.

MEASURING IMPACTS OF SCIENCE, TECHNOLOGY, AND INNOVATION INVESTMENTS IN BRAZIL

Brazil's Marcio de Miranda Santos, Executive Director of the Center for Strategic Management and Studies in Science, Technology, and Innovation, explained why quality data and a good information gathering system are invaluable for evaluating research impacts and outcomes. A comprehensive information infrastructure that facilitates evaluation of research is difficult to build, since many types of information are necessary for a thorough evaluation, including data on individual researchers, projects, collaborations, R and D networks, research institutions, and public agencies. The system has to be adaptable and able to handle the complexity of a range of inputs. Santos described Brazil's strategy for building such a system.

Several principles are guiding the center's work. One is to expand on what is already available. In Brazil, this means linking data from sources such as the National Council for Science and Technology, the National Agency for Industrial Development, various innovation agencies, projects, and dissertations. The data requirements must be designed not just for government needs but to provide access and functionality for science, technology, and innovation participants as well. An effective program will rely on traditional software engineering methods as well as knowledge engineering and e-government approaches.

The Lattes platform, which Brazil has been using since 1999, holds program information in a database that currently contains over 2 million curricula vitae (CVs) and is updated every three months on average. In 2008 the Center used it successfully to do an ex-ante evaluation of networks that had submitted proposals to the National Institutes of Science and Technology program (INCT). The program aims to promote networks among research groups and individuals, internationally competitive research, high-quality S and T development, and joint use of laboratories by universities and companies. The program also will contribute to improving education standards at all levels.

Using Lattes, the Center took snapshots of information from individual CVs and from the INCT program as a whole and analyzed that information to determine the success of the program. A snapshot of one project from 2008, with 25 people in the network, provided data on co-authorship of papers, researchers who shared advisors, and participation in other projects and committees. The Center then used Innovation

Portal, an electronic service designed to link information from different data sources, to follow shifts in project networks and collaboration. For example, three scientists working on the first project were not co-authors at the time the proposal was submitted, but by 2011 they had begun to produce papers with other project participants.

Another example comes from the Brazilian Academy of Mechanical Sciences and Engineering, which was interested in identifying the weaker departments in mechanical engineering in Brazil. Researchers used Lattes to examine the distribution of knowledge within mechanical engineering, based on the number of publications produced by each scientific domain. They broke down the field into smaller subdomains and pinpointed weaker areas where reinforcement would be useful. This methodology allowed public decision makers to not only identify weak spots but also track improvement, measure the impact of research investments, and make decisions on how to further improve the system.

The advantages of an integrated national platform such as Lattes are substantial, said Santos. It allows efficiency in both ex-ante and ex-post evaluation processes, increased transparency, and increased community participation. Research institutions, individuals, and firms are able to access the Lattes platform as well, so it is an open system not limited to the government, and groups become aware of their own progress and that of other teams and programs. Some areas are still weak, but the center is currently developing a system to incorporate more information from the private sector in particular, which is one of the largest gaps.

"[The platform] facilitates the participation of the scientific community," said Santos. "If the scientific community knows what's going on, it will be better for national federal agencies to interact and allow for the community to participate, because they know they have access to information."

DISCUSSION

During the question period a participant asked Viney how the U.K. Medical Research Council (MRC) convinces grantees to participate in the e-Val system, since it is more time consuming than end-of-grant reporting. Viney explained that the MRC has been successful at getting increased government funding using data from the e-Val, which they can use to leverage participation since the research community is able to see the impact of providing such detailed reports. The e-Val is also mandatory for new grants, so participants must comply if they want to receive MRC funding in the future.

Responding to questions about how the impact on policy is measured, Sloan explained that the Framework Program has attempted to study the impact of their projects on policymaking by questioning participants, but has not done citation analysis of policy documents. Viney said that the MRC has looked at where MRC research is cited, paying particular attention to which documents are more influential and tracking any resulting policy changes.

A workshop participant asked about the European Commission's guarantee fund, where some money is held back until participants fulfill the requirements of the grant, and whether surveys must be completed in order to receive that money. Sloan said that it depends on how strongly the requirements are enforced, but that much of what is asked is voluntary.

In response to a question asking whether a clear policy is in place requiring researchers to acknowledge their funding when they publish a paper or develop a patent, Viney said that analysis of citations and publications is based on the most reliable data possible. He said that research councils in the UK do require a standard type of acknowledgement in publications, but that the MRC could potentially do a better job working with publishers and checking compliance. Santos added that in Brazil, there are policies for federal agency funding and some state funding, but there is room for improvement so that their system is able to capture exactly who funded what.

8

EMERGING METRICS AND MODELS

Continuing progress in measuring the returns on research investments requires new metrics and models to analyze how the inputs to research are converted into both short-term outputs and long-term impacts. NSF Director Subra Suresh provided the context for this discussion in a lunchtime keynote address that described five themes guiding NSF's investment decisions. Two separate sessions at the workshop included seven speakers who examined specific tools and approaches, from the creation of a science policy infrastructure at NSF to visual analytics that can probe data sets for unexpected findings.

ASSESSING RESEARCH AT NSF

Traditional measures of research outputs provide only a partial picture of the state of scientific research in the United States, said NSF Director Subra Suresh during his keynote address at the workshop. For example, if the percentage of scientific publications were extrapolated into the future based on the trends of the last few years, China's percentage would surpass that of the United States in 2013 or 2014. Publications are only one metric, Suresh acknowledged, and their impact is a matter of debate, but "agencies like NSF are looking at the significance, or lack thereof, of these kinds of metrics."

Taking a different metric, the United States led the world until 2000 in R and D expenditures as a fraction of GDP. But in that year three major competitors —Germany, Japan, and South Korea— surpassed the United States, and several smaller countries have done so since. Other countries, such as China and Singapore, are investing very heavily in science and engineering research.

69

With the increasing globalization of research, metrics of the United States' competitive edge will inevitably change. But such changes raise the question, said Suresh, of "what kind of metrics do we put in place so that we can position ourselves most appropriately for the future?"

At the National Science Foundation, this question should be considered within the context of five broad themes that are guiding the agency. First, science has entered what Suresh called a "new era of observation." Digital technologies make it possible to generate data at an unprecedented pace. These data, along with new computational tools, are creating both tremendous excitement and new problems. NSF is devoting considerable effort to the development of cyberinfrastructure that can take advantage of these opportunities and solve the problems. In particular, cyberinfrastructure provides new capabilities for assessment of research. For example, the agency is asking what kinds of capabilities it can put in place in situations where the research community uploads data and information automatically. Researchers already have many responsibilities, and NSF has to be careful not to impose unfunded mandates on the community, said Suresh. But cyberinfrastructure makes it possible to store, integrate, sort, extract, and permanently archive information. How can this information best be used while protecting the integrity and confidentiality of the scientific process, Suresh asked. How can NSF work with other federal agencies and with its counterparts around the world to use this information to move science and education forward?

A second important opportunity, according to Suresh, is to integrate data and ideas from the social sciences and from the natural sciences. As an example, Suresh described NSF-sponsored research that identified the potential economic benefits of auctioning off portions of the electromagnetic spectrum. The 2012 federal budget projected that such auctions are expected to yield approximately $28 billion over the next decade, with $10 billion of that being set aside budget deficit reduction. "That's a tangible contribution to policy of social sciences research sponsored by NSF some 20 years ago," Suresh said. The social sciences research being sponsored by NSF offers many similar opportunities to leverage natural sciences research. In the context of clean energy, for example, Suresh has been talking with officials at the Department of Energy on how social, behavioral, and economic research sponsored by NSF can contribute to research supported by the department.

A third opportunity is to expand research partnerships both within the United States and internationally and through people exchanges as well as virtually through digital technologies. As NSF lacks the capability to engage in multiple bilateral relationships with many countries, Suresh has been exploring how NSF can work with private foundations and with multilateral bodies such as the G20 countries to enhance international cooperation.

Suresh's fourth theme was the need to continue investing in the development of human capital, especially the STEM workforce, not just for the United States but for the world. Since 1952, Suresh noted, NSF has funded 46,000 graduate research fellows. In 2010 it doubled the number of graduate fellows to 2,000 per year and kept the number at 2,000 in 2011. In addition, the stipend was increased from $10,500 to $12,000, and NSF's goal is to sustain that level of support into the future. NSF's' initial graduate fellows would be well into retirement by now. How were their careers shaped by NSF's support? Have the fellowships helped women and underrepresented minority groups over the past 58 years? What effect have career awards and young investigator awards had on researchers? New computer technologies could gather information to help answer some of these questions and shape human capital policies within the financial constraints expected in the future."

A fifth theme was the need to measure the impacts of NSF funded research intelligently and over a long period of time. Although a good deal of the research NSF funds has purely scientific motivations, some of it has helped generate entirely new industries making significant contributions to the economy, Suresh observed. How can NSF help match the products of research with the needs of the marketplace without taking money away from fundamental research? How can the agency reconcile the short-term economic focus of the country and its elected leaders with the long-term benefits of basic research? How can NSF best articulate the benefits of basic research funding over the course of decades for the American public and the global society? Suresh suggested that a possible model could be the studies of higher education institutions' contributions to the economy of the Boston area. He also cited the number of startup companies that have emerged in part from NSF-funded nanoscience and engineering centers. In addition, he recounted physicist Michael Faraday's response to William Gladstone when asked about the practical value of electricity. Faraday replied, "One day, sir, you may tax it."

Suresh concluded his remarks with an invitation to workshop participants to make suggestions to NSF on its policies and programs: What new kinds of programs need to be put in place to take advantage of current opportunities? Should NSF's merit review process be changed to recognize truly transformative multidisciplinary research? Can NSF promote family-friendly policies that will enable women in much greater numbers to join STEM workforce? Such input "would be enormously helpful," Suresh said.

THE STAR METRICS PROJECT

In 2005, OSTP Director John Marburger observed at a AAAS policy forum that he found it very difficult to provide an evidence-based answer to the question, "How can the federal government optimize its investments in science?" An interagency working group under the title of Science of Science Policy came to a similar conclusion in 2008, noting that no solid theoretical and empirical basis exists for deciding the level or allocation of scientific investments.

Those observations, along with the establishment of the Science of Science and Innovation Policy (SciSIP) program at NSF, culminated in an initiative to build a data infrastructure that would help answer the questions posed by Marburger and the interagency group. SciSIP Director Julia Lane described this system, known as STAR Metrics, at the workshop.

The Motivation for STAR Metrics

The motivation behind the system is threefold, said Lane. First, a principle of good government is that officials should be able to document the results of government spending. Instead, she said, most agencies are unable to document what researchers are supported, let alone what are the results of their work. Second, agencies need to be responsive to stakeholders, and the Office of Management and Budget, Office of Science and Technology Policy, and Congress are all asking for data. Third, the utility of the data requires new analytical approaches and the use of cutting edge technologies. "Relying on manual and burdensome reporting simply doesn't make sense."

What is STAR Metrics?

STAR Metrics is a federal and university partnership to document the outcomes of science investments to the public. It is an OSTP initiative partnering with NIH, NSF, DOE, and EPA that is divided into two phases. Phase 1 involves establishing uniform, auditable, and standardized measures of the initial impact of ARRA and base budget science spending on job creation. Phase II calls for the collaborative development of measures of the impact of federal science investments on the creation and diffusion of scientific knowledge (through publications and citations), economic growth (through patents, start-ups, and other measures), workforce development (through student mobility and employment), and social outcomes such as health and the environment.

This represents what Lane termed a "sea change" from the current data infrastructure on public science. For 50 years, the science agencies have essentially been proposal processing and award administration factories, she said. They apply labor and capital to the receipt of proposals, the awarding of grants and contracts, and the management of their performance. The proposal or award is not a behavioral unit of analysis but an intervention. The behavioral unit of analysis is the individual scientist. There is a pressing need, said Lane, is to restructure the data system to "look at the human beings who are affected by science funding and try to explain their behavior."

Nevertheless, observed Lane, it makes less and less sense to talk about the outcome of an individual award. Increasingly, the relevant unit of analysis is a cluster of researchers, a scientific field or subdiscipline, or an entire research agenda. In addition, principal investigators typically get funding from a stream of activities, so being able to identify the incremental impact of an individual award is extraordinarily difficult. This has implications for the structure of the data within the agencies. "You have to capture the activities of the scientists over their entire period of activity, not just the period of the award." Finally, the outcomes of many awards occur long after the administration of the award. Unless this long-term benefit is measured, the impact of a scientific investment will be under-estimated.

Capturing Data

In the twenty-first century, almost all scientific activity occurs electronically, yet reporting of scientific activities is often still done manually. "Submitting data that are in PDF format that are unstructured and unsearchable means that you miss enormous amounts of what's going on," said Lane.

In phase I, the STAR Metrics program sought to capture who is being supported by scientific funding without burdening researchers. It did that by using the internal administrative records of researchers' institutions to capture that information as it flows from one place to another. STAR Metrics receives 14 administrative data elements from awards, grants, human resources, or finance systems on a quarterly basis.

Phase I began with a pilot project at six institutions. Since then, 75 institutions have joined on a voluntary basis. The data need not be personally identifiable.

As an example of the information that can be generated in phase I, Lane cited data on full time equivalent (FTE) positions. The data yield quarterly reports on FTE jobs generated by ARRA, total FTE jobs and positions, FTE jobs generated through subawards and among vendors, and jobs generated through overhead payments. "For the first time, for each institution, we're able to document how many people are supported," Lane said. Faculty are only a small proportion - about 20 percent - of the FTEs that are supported. Support services, graduate students, postdoctoral fellows, undergraduate students, and others represent 80 percent of the supported positions. An FTE may represent several supported students. The data also make it possible to calculate the total number of individuals supported by research funding, along with the number of positions supported outside universities through vendor and subcontractor funding. "Not a single PI lifted a pen or typed a keyboard to enable us to pull this information, yet the information is very powerful and can be used to inform federal and state lawmakers."

Future Plans

The next step in STAR Metrics' development is to develop the main features of the phase II platform that will compile information from individual researchers, commercial publication databases, administrative data, and other sources to capture as much information about scientific activities as possible. Federal policymakers, agency officials, research institutions, and investigators "will have a common and coherent system of understanding what they're doing and the impact of what they're doing," Lane said.

RECONSTRUCTING NETWORKS OF DISCOVERY

The media have been questioning the return on federal research investments, noted Stefano Bertuzzi from the Office of Science Policy Analysis in NIH's Office of the Director. A 2008 article in *Newsweek* concluded that "judging by the only criterion that matters to patients and taxpayers— not how many interesting discoveries have been made, but how many treatments for disease the money has bought— the return on investment to the American taxpayer has been approximately as satisfying as the AIG bailout." A more recent article in *Nature* entitled "What Science Is Really Worth" ran under the tagline, "Spending on science is one of the best ways to generate jobs and economic growth, say research advocates. But the evidence behind such claims is patchy."

Building an Empirical Framework

Continuing the discussion of STAR Metrics, Bertuzzi described it as a way of combining and linking input measures with economic, scientific, and social outcomes. For example, when a new discovery or technology is licensed to a company, the license represents a return on research investments. STAR Msfrics would "unpack what is inside the black box of the licensing," said Bertuzzi.

Bertuzzi demonstrated a prototype tool based on the discovery of drugs for rheumatoid disease. These are transformative drugs that can seem to bring people back from near death, and they generate billions of dollars in sales each year. Using information from STAR Metrics, it is possible to trace the developments that led to these drugs using the scientist as the unit of analysis.

The scientific story began with fundamental research on inflammation, which led to the discovery of tumor necrosis factor (TNF). Further research on molecular mechanisms involving TNF gave rise to several different drugs that work in different ways to reduce inflammation.

STAR Metrics data show the levels of public and private funding for this research as based on funding attributions in publications related to TNF. Funding began largely in the public sector at NIH and then decreased over time as private funding increased. The data also yield an interactive website that presents a timeline of milestone events that led to the approval of specific drugs. Clicking on an event in the timeline produces a list of the scientists involved in publishing key papers. Clicking on the paper pulls up a brief CV along with highlights of the discovery and funding sources. Further links connect scientists with patent databases and other information.

The links among scientists, discoveries, publications, patents, and other information form networks that allow the process of discovery to be visualized. Interactive websites make it possible to explore the network to uncover collaborations, institutional connections, linked events, and other aspects of innovation. "We will be able to collect, through federal-wide profiles, what the scientists themselves tell about their stories, their interests, and their discoveries," said Bertuzzi. STAR Metrics will make it possible to "disentangle and unpack all the complexity of the network that eventually led to that particular discovery." A potential practical application would be to look for the common features of successful discovery processes and then try to replicate them.

CREATING KNOWLEDGE FROM DATA

The outputs of research historically have been viewed as consisting of papers, patents and human resources, noted Ian Foster, Arthur Holly Compton Distinguished Service Professor and Chan Soon-Shiong Scholar at the University of Chicago. Papers document ideas, patents establish ownership rights, and human resources constitute people who are trained in ideas and in methods.

Today, said Foster, large amounts of human intellectual capital are being captured in other forms— especially as data and computer software. These resources also capture ideas and methods that can be transferred from one person to another. Such resources have been growing explosively. In 2001, according to an annual report from the journal *Nucleic Acids Research* on the number of publicly available, high-quality databases in molecular biology, there were 96 molecular biology databases. In 2010, there were 1,070, and in 2011 there were 1,330. Some of these databases have tens of millions of entries and billions of bytes of nucleic acid information. "Historically, we might have thought of people as conducting an experiment, writing it up, and putting the results into a paper which other people would read, build on, and perhaps cite in their publications. Clearly, consulting databases rather than the literature has become a primary means of accessing the work of other investigators."

In addition, an expanding set of online services provide access to software. "Web services" is a term often used to refer to the software that is made available over the internet by standardized protocols. One registry lists 2,053 services provided by 148 providers. Some of these provide very simple functions, but others provide sophisticated computational capabilities to scientists who otherwise would not have access to them. Furthermore, many of these services are made freely available to others, often through large development and distribution communities. "Data and software are two types of resources that are becoming fundamental to how people do science, and they are being shared in ways that are very different than just a few years ago."

New methods are needed for evaluating these resources, said Foster, including their impact on the research process as well as on downstream activities such as job creation, patenting, and the formation of companies. The fact that these resources are digital makes such evaluations somewhat easier, because accessing an electronic database or piece of software involves a digitally mediated transaction and can be logged and analyzed in the future. Collective analysis of these transactions, along with more conventional metrics, also can reveal the ways in which knowledge is integrated. For example, the MyExperiment project seeks to make the sharing of computational procedures, data, and software as easy as sharing images on a social networking site. The site also makes it possible to share workflows and reports on how often they are used and for what purpose. "We can look not only at how people interact with people via publications but also how software interacts with data and data with software and people with software and data."

The STAR Metrics program also seeks to capture research activities and outputs in the form of a distributed database. In that context, it becomes possible to automate many administrative tasks such as creating biosketches, progress reports, final reports, and tenure reviews.

In this and other ways, researchers derive tremendous value from such platforms, said Foster. Researchers are as interested as evaluators in the connections between different knowledge bases. A system that links all research outputs to all relevant research inputs would be invaluable to researchers who are trying to determine which pathways have not been explored and should be pursued, which research strategies are most useful, and how a particular research problem has been tackled in the past. "With luck we will find, as is often the case in science, that the very activity of observing something will change the activity that we are observing, and accelerate its process."

MEASURING THE IMPACT OF STAR SCIENTISTS

Measuring the impact of research requires a long-term view, said Lynne Zucker, Professor of Sociology and Policy Studies at the University of California, Los Angeles. The short-term impact can be much smaller than the long-run impact. To see these long-term impacts, said Zucker, "ten years out is about the minimum, in my experience, from having done a lot of evaluations of programs for the University of California system and for the Advanced Technology Program and other programs."

Many new ideas are embodied in those who conceive them. People have high amounts of tacit knowledge, and they can transmit this knowledge to others. People who have been doing the same kind of science often can absorb these ideas quickly, but in general the diffusion of ideas is slow. Teams that include what Zucker called "star scientists" have been located primarily in universities, but increasingly they occur in firms, too. "There's a lot of basic science going on in industry," said Zucker.

Biotechnology is an exemplar of a science-driven industry. Scientific breakthroughs led to hundreds of new firms. Consolidation occurred when scientific advances slowed, with some firms growing and others failing. However, the number of jobs continued to grow, so that people were absorbed into the successful companies. In the case of biotechnology, the growth and change were revolutionary enough that an entirely new industry was created.

Developing an infrastructure to collect data about knowledge flows into industry is a complicated process and has not been done well in most industries, according to Zucker. However, in biotechnology, a system known as Bioscan makes it possible to track the process of transferring knowledge from molecular biology into industry. Bioscan also shows that firms in which star scientists are involved have higher employment growth than others. "It's a selection process— the top talent gets selected first," said Zucker

A new model of a high-science firm emerged in biotechnology. Scientists were free to publish and were rewarded for it, both in salary and stock options. Firms had deep collaborations with university faculty, and rewards were closely tied to the firms' outputs. Large incumbent firms learned to emulate this culture, and if they did not they had a tendency to fade and die.

More recently, many nanotechnology firms have been adopting the biotech model and are undergoing a similar process. Many startup and incumbent firms are competing, with roughly one in ten firms having star scientists involved in their firms. Nanotechnology is more geographically distributed in the United States than biotechnology. But where star nano-scientists are active has been a key determinant of where and when new firms enter the field.

NSF funding for nanotechnology has had a large impact in the field, Zucker observed, contributing to large increases in published nanoscale articles and significant growth in nanoscale patenting.

The impacts of star scientists vary across S and T areas in proportion to technological opportunity, said Zucker. Some areas have had recent breakthroughs, and those areas are going to have more opportunities than areas where the science is more mature. But scientific fields also make their own opportunities, as when biotechnology firms have begun working in nanotechnology.

In general, said Zucker, federal investments appear to be important for impacts in all science and technology areas, but to test this idea she and her colleagues have been developing an integrated database with input from multiple sources. The resource is beginning to produce early results, and "the general answer so far is yes, with some variation, federal grants do make a big difference . . . for most science areas."

The initial version of the resource, StarTechZD, is now available on the web (http://startechzd.net) and permits the tracking of knowledge, funding, and economic impacts. It can identify both organizations and particular scientists within and across databases. It also can separate organizational and individual efforts. Zucker called it a "quantum jump in the ability to analyze science and technology. . . It's an extremely important tool."

VISUAL ANALYTICS

Visual analytics is the science of analytical reasoning facilitated by interactive visual interfaces, said John Stasko, Professor and Associate Chair of the School of Interactive Computing at the Georgia Institute of Technology. It combines automated analysis techniques with interactive visualizations for effective understanding, reasoning, and decision making on the basis of large and complex data sets. Another way to think of visual analytics, said Stasko, is that it combines interactive visualization, computational data analysis, and analytical reasoning. "Visualization is not about making pretty pictures," he said. "It's about helping people solve problems and gain insights from their data."

Visualization is not appropriate for every problem. If someone is interested in how many people are employed in an area, a data mining algorithm can find the best fit. However, visualization is a powerful tool in exploratory data analysis scenarios, "where someone drops a pile of data in your lap and says 'Help me understand what's there.'" These are scenarios were people typically do not know exactly which questions to ask.

Effective visualization tools both answer questions and raise questions. The interactive aspects of the data enable someone using to tool to essentially have a conversation with the data. "You explore one angle and a new question arises. It's through the interaction where things happen."

Some existing visualizations can be frustrating, Stasko admitted. For example, large network graphs such as maps of science do not necessarily convey clear conclusions. A map might show that mathematics is strongly related to computer science, but such an observation is not very interesting. Also, one visualization cannot necessarily show all of the variables that someone might want to represent. They present a static view of connectivity, clustering, or centrality, "but you want to go beyond that."

Stasko cited several examples of effective interactive visualizations. The Social Action system uses social network analysis to measure the centrality of different nodes in the network, thus combining the algorithmic analysis of the data with interactive exploration. Another system called Jigsaw does document analysis of unstructured text. Through such processes as text mining and entity identification, it produces multiple interactive visualizations of the content of the documents for exploration. Finally, Stasko mentioned a system called Ploceus (named after a weaver bird that creates elaborate nests) that does network visualizations from tabular data. The system takes data from a spreadsheet, for example, and creates networks that allow the data to be explored.

Stasko concluded by saying that there are many different methods of data analysis and they are not mutually exclusive. The best kinds of data analysis combine statistical, automated computational, and visual exploratory methods, he said. From such explorations of data, where the questions are not necessarily defined beforehand, insightful discoveries can emerge.

CONSIDERATIONS IN BUILDING COMPREHENSIVE DATABASES

Adam Jaffe, Dean of Arts and Sciences and Fred C. Hecht Professor in Economic at Brandeis University, commented on the importance of creating a comprehensive database that contains all research inputs and outputs. "It has been a long time in coming, and we've talked about it for a long time, but we are now at a point where we can glimpse that it may actually be happening." The only thing that can protect science funding, he said, is demonstrating the long-term and diffuse but tremendously important impacts of science, "and that requires very extensive and complicated data."

One way to build such a database will be to take advantage of automated data capture. Once the framework for the system has been created, huge amounts of data can be collected automatically by searching the web. Automated data capture will reduce the reporting obligations imposed on institutions and individuals. "The ARRA reporting requirements almost caused my office for research administration to implode," said Jaffe. Universities are under stress because financial support from all sources is down while financial needs are up. "Everyone is overworked, and when you put these reporting requirements on top of that, it really is a significant issue that we need to worry about."

Such a database would be greatly advanced by a unique identifier for each person who receives money from the federal government to conduct research. "This is absolutely crucial," said Jaffe. "If we eventually fail to get to a system where each person is tagged with a unique identifier, this project will not succeed." Real data have many ambiguities that need to be resolved, and a unique identifier would resolve many of them.

Evaluations also need to track the failures—the students who dropped out, the grant applications that were not funded, the projects that produced negative results. "You don't know the return to the successful investments unless you can have some kind of 'but for' or counterfactual to compare what occurred when you funded it to what might have occurred otherwise." Statistically, the best way to answer these questions is to have data in the system on other than successful outcomes.

Finally, Jaffe said, the data should extend beyond the biosciences. "I know NIH is the 800-pound funding gorilla, but there are other sciences and other industries out there."

The indirect effects of research funding can be very difficult to track. Things like the accumulation of human capital or the spillover effects from research have very long lags and diffuse impacts. Data collection therefore needs to be broad-based and multidimensional. "What is so exciting about some of these projects is that we are beginning to see an infrastructure where all the different pieces can be connected together, where we can come to understand better how all these things work."

DISCUSSION

During the discussion period, the panelists discussed several prominent issues associated with improving the accuracy of information in databases. Administrative data tend to contain many errors, which can reduce the value of analyses. Some disciplines have adopted systems in which researchers are asked to review and correct errors in, for example, listings of publications and citations. One approach would be to promote researchers' retention of permanent e-mail addresses that could function both as identifiers and as a means of verifying information related to that person.

Julia Lane cautioned that a unique identifier for each researcher may not be practical and may not be essential. It may make more sense to think of investigators having multiple identifiers that are interoperable. Identification is a problem in many countries, not just the United States, and efforts both within and across nations are now reaching the point where progress can be made.

Spector suggested that databases need to leverage the federated transparency of the Web rather than creating specific systems for measuring the impacts of research. There are several ways of doing this. Crowd-sourcing can be "incredibly powerful" because many people, and particularly the younger generation, want to keep information up to date. Natural language processing can help improve accuracy by comparing information from many places on the Web. Finally, machine learning algorithms are powerful categorization mechanisms. "Don't build custom systems," Spector warned, "because they will be expensive [and] bureaucratic."

In response to a question about how advances in data presentation and visualization can help policymakers better understand and use data, Stasko said that it is critical for the designers of such systems to understand the systems' users and tasks. "What do you want to find out about the data, and how can visualizations help?" The answers to questions in areas such as patenting could change scientific practices and help set the research agenda. And visualization can help convey the complexity of the innovation ecosystem, with all its different and tangled components.

Director Suresh was asked about the "broader impacts" criterion that NSF uses to review proposals, with reference to the reauthorization of the America COMPETES Act calling on NSF to broaden these impacts to include such considerations as performance measures and partnerships. Suresh responded that the National Science Board has been investigating the broader impacts criterion. Researchers are understandably confused, he said, about how many of these considerations to incorporate into their research proposals, how much of the burden to place on the individual versus the department versus the school versus the institution, and how to consider such factors as economic impact and workforce development. "This is very much a work in progress." A number of groups are working in parallel and in conversation with one another, he said, ideally leading to clarity rather than confusion on this issue.

9

PITFALLS, PROGRESS, AND OPPORTUNITIES

During the final session of the meeting, members of a panel shared their opinions of the major messages and unanswered questions that emerged from the two days of presentations and discussions. Theirs were individual observations rather than an expression of a consensus on the part of the panel or the workshop participants as a whole.

PITFALLS ON THE ROAD TO UNDERSTANDING

- The selection of specific measures inevitably focuses attention and effort on what is being measured. Their value can decay as more of what is being measured is produced. Further, the selection of metrics can reduce the valuable diversity of the research system and its potential for serendipity.

- Agricultural research has been so successful that Americans now pay less for food per capita than in almost any other country and any other time. This success may have had the perverse effect of undermining funding for basic agricultural science, since the need for productivity gains seems less pressing.

- Research funding volatility has major consequences for the decisions made by research performers. For example, the doubling of the NIH budget drove a large expansion of biomedical research facilities at research universities in the expectation that increases would continue. The suspension of real growth at NIH halted the growth of indirect cost recovery to pay for those buildings, with adverse effects for other parts of the university. Funding patterns also send messages to students about desirable fields of research – messages that may be at odds with long-term employment prospects in those fields. Volatility is problematic in firms as well as in federal research.

- An assessment of even a narrow field requires taking an average from disparate processes and systems, which can cause such assessments to be overly broad. For example, the number of patents granted within a particular field may be important, but individual researchers should not be judged by how many patents each one has generated.

- If all past research had been required to justify its value in terms of practical benefits, advances that have led to massive practical benefits would not have occurred.

- The knowledge generated by fundamental research has an intrinsic value regardless of its application. Without it, applied work would stagnate.

- Policymakers and the public in general agree on the value of research. Could research that fails to identify many of the benefits of science undermine that consensus and therefore be harmful?

PROGRESS IN UNDERSTANDING THE ISSUES

- The ever-growing power of the Web and the information sharing it enables will facilitate the analysis of research outputs. Natural language processing, machine learning technologies, and crowd sourcing will increasingly glean many reasonably accurate metrics from publications, patents, social networks, blogs, and so forth, and this capability will increase over time. Furthermore, this approach will be less costly and provide more information than government-mandated reporting. However, government agencies will need to create new tools to use these data to help fulfill their missions.

- The benefits of research results, both in terms of new knowledge and trained students, are vastly different from discipline to discipline and even from subdiscipline to subdiscipline. Thus, the determination of impacts requires very detailed analysis that is highly sector specific. For example, the evaluation of physics is different than the evaluation of computer science, and the evaluation of theoretical computer science is different than the evaluation of research in parallel computation.

- Relatively small expenditures on increasing the dissemination of research results could greatly enhance the beneficial impacts of research without entailing major new funding initiatives.

- Some questions may not be answerable, and identifying those questions may usefully focus attention on the questions that can be answered.

- The major discoveries that result from basic research are outliers that generally are very hard to predict. They emerge rarely, but they are the most important. How can these be accommodated in assessments of the value of research?

- As science becomes more interdisciplinary, more collaborative, more international, more digital, more open, more expensive, more diverse, and more fast-paced, measuring impacts will face new and difficult challenges.

OPPORTUNITIES POSED BY GREATER UNDERSTANDING

- The science of science policy has an opportunity to examine the broader issues of economic growth and societal change if it interprets its agenda broadly. As an example of an important albeit difficult question, are additional funds most usefully spent on health-related R and D or on insurance? Some analysts have cited the drop in deaths from cardiovascular disease starting in 1965 as an outcome of biomedical research, but that was also the year when Medicare was instituted.

- The plural of anecdotes may not be data, but anecdotes can be more powerful than data in swaying policymakers, even if they are not necessarily representative.

- A heightened emphasis on accountability within government will increase the need to produce metrics of research impacts. The research community needs to understand why this is important, especially because they can contribute ideas that would benefit data collection and analysis.

- Research funders and performers have many opportunities to work with the private sector in measuring the impacts of research, since the private sector spends considerable time and money working on this issue.

- The optimal amount of research for the United States as a percentage of GDP still has not been determined. Is it possible to overspend on R and D? To what extent should education be emphasized in that spending?

CONCLUDING REMARKS

Finally, several speakers on the panel emphasized that scholars studying these issues should be humble, sensitive, and do no harm, which is a message Irwin Feller delivered at the beginning of the workshop. The returns on federal investments in research are extremely complex and occur within the context of a complex economy and society. Analysts should avoid claiming more for the utility of their work than is warranted.

REFERENCES

Bozeman, B., and D. Sarewitz. 2005. Public values and public failure in US science policy. *Science and Public Policy,* 32:119-136.

Committee on Prospering in the Global Economy of the 21st Century. 2007. *Rising Above the Gathering Storm: Energizing and Employing America for a Brighter Economic Future.* Washington, DC: National Academies Press.

Cutler, D., and Kadiyala, S. 2003. The returns to biomedical research: treatment and behavioral effects. In *Measuring the Gains from Medical Research: An Economic Approach,* edited by K Murphy and R. Topel, pp. 110-162. Chicago: University of Chicago Press.

Health Economics Research Group, Office of Health Economics, and RAND Europe. 2008. *Medical Research: What's It worth? Estimating the Economic Benefits from Medical Research in the UK.* London: Evaluation Forum.

Heidenreich, P., and McClellan, M. 2003. Biomedical research and then some: the causes of technological change in heart attack treatment. In *Measuring the Gains from Medical Research: An Economic Approach,* edited by K. Murphy and R. Topel, pp. 163-205. Chicago: University of Chicago Press.

Lichtenberg, F. R., and Sampat, B. N. 2011. What are the respective roles of the public and private sectors in pharmaceutical innovation? *Health Affairs* 30(2):332-9.

Manton, K., Gu, X., Lowrimore, G., Ullian, A., and Tolley, H. 2009. NIH funding trajectories and their correlations with US health dynamics from 1950 to 2004. *Proceedings of the National Academy of Science USA,* 106(27): 10981-6.

Morlacchi, P., and Nelson, R. 2011. How medical practice evolves: the case of the left ventricular assist device. *Research Policy,* 40(4):511-525

Pardey, P. G., Alston, J. M., Christian, J. E., and Fan, S. 1996. *Hidden Harvest: U.S. Benefits from International Research Aid.* Washington, DC: International Food Policy Research Institute.

Roach, M., and Cohen, W. 2011. Patent citations as measures of knowledge flows from public research: a comparison with survey data from U.S. R and D labs. In preparation.

Roach, M., and Sauermann, H. 2010. A taste for science? Ph.D. scientists' academic orientation and self-selection into research careers in industry. *Research Policy*, 39(3):422-434.

Sampat, B. N., and Azoulay, P. 2011. The impact of publicly funded biomedical and health research: a review. (See Appendix D)

Stevens, A. J., Jensen, J.J., Wyller, K., Kilgore, P.C., Chatterjee, S. and Rohrbaugh, M.. 2011. The role of public-sector research in the discovery of drugs and vaccines. *The New England Journal of Medicine,* 364(6):535-541.

Toole, A. A. 2007. Does public scientific research complement private investment in research and development in the pharmaceutical industry? *The Journal of Law and Economics,* 50(1):81-104.

Ward, M., and Dranove, D. 1995. The vertical chain of research and development in the pharmaceutical industry. *Economic Inquiry,* 33: 70-87.

APPENDIX A

WORKSHOP AGENDA

Measuring the Impacts of Federal Investments in Research
April 18-19, 2011
20 F Street N.W. Conference Center

Washington, D.C. 20001
APRIL 18, 2011

7: 30 AM Registration

8: 15 AM Introductions, and Workshop Objectives

Neal Lane, Co-Chair; University Professor, Rice University
Bronwyn Hall, Co-Chair; Professor, University of California, Berkeley
and University of Maastricht

8:30 AM Welcome Address: The Honorable Rush Holt (D-NJ), U.S. House of Representatives

Introduced by: **Neal Lane**, Co-Chair

8:45 AM Session I: Promise and Limits of Measuring the Impact of Federally Supported Research

What have we learned from previous efforts to measure the economic impact of federal research investments? What approaches and metrics are more and less promising? What are the noneconomic factors that could be used as alternative measures of the impact of federal research?[1]

Moderator: **Alfred Spector**, Vice-President, Google, Inc.
Commissioned Paper Presentation: **Irwin Feller**, Professor Emeritus, Economics, Pennsylvania State University
Commentator: **Daniel Sarewitz**, Professor of Science and Society, Arizona State University

9: 30 AM Discussion

10:00 AM Break

10:15 AM Session II: Aggregate Impact of Federally-Supported Research on the U.S. Economy and Quality of Life

What do we know about or how could we determine the contributions of public research to: GDP and productivity? Wages and employment? Private sector R and D and innovation? Is there any basis for setting a target for aggregate research expenditures?

Moderator: **Bronwyn Hall**, Co-Chair
Panelists:
Carol Corrado, Senior Advisor and Research Director in Economics, The Conference Board
Bruce Weinberg, Professor of Economics and Public Administration, Ohio State University
Michael Roach, Assistant Professor of Strategy and Entrepreneurship, Kenan-Flagler Business School, University of North Carolina

11:15 AM Discussion

11:45 AM Lunch Break

---

[1] The questions listed for each session of the workshop were intended to stimulate thought and discussion. It was not expected that presenters would address all of these questions nor that the session as a whole would provide the answers.

12:30 PM Session III: Funding and Impact of Biomedical and Health Research

What are the links between publicly funded research, biomedical innovation, and health outcomes and costs? Are there metrics that could help policymakers strengthen those linkages? What have we learned about the effects of fluctuations in the National Institutes of Health funding over the past decade and how to manage future funding changes? How do private firms and philanthropic organizations gauge the results of their health-related research investments?

Moderator: **Neal Lane**, Co-Chair
Commissioned Paper Presentation: **Bhaven Sampat**, Assistant Professor of Public Health, Columbia University
Panelists:
Richard Freeman, Herbert Ascherman Chair in Economics, Harvard University
Paul Citron, Retired Vice-President, Technology Policy and Academic Relations, Medtronic, Inc
Laura Guay, Vice-President of Research, The Elizabeth Glaser Pediatric AIDS Foundation

2: 00 PM Discussion

2:30 PM Break

2:45 PM Session IV: International Perspectives on Assessing Research Impacts

What progress has been made abroad in tracking and assessing public research outcomes? What methods and metrics might be applicable in the United States? What features of national research systems make it easier or more difficult to transfer methodologies?

Moderated by: **Bronwyn Hall**, Co-Chair
Panelists:
Ian Viney, Head of Evaluation, Strategy Group, Medical Research Council, United Kingdom
Brian Sloan, Directorate-General, Research and Innovation, European Commission

Marcio de Miranda Santos, Executive Director, Centre for Strategic Management and Studies in Science, Technology and Innovation, Brazil

4:00 PM Session V: Assessing Mixed Market and Non-Market Impacts of Research

Can we measure the less-quantifiable benefits of research such as on climate change mitigation, food security, environmental protection, and national security? What are the alternative approaches for better assessing the non-market impacts of research? How do private firms and foundations measure the results of their research investments related to public goods?

Moderator: **Catherine Woteki**, Under Secretary for Agriculture for Research, Education and Economics, U.S. Department of Agriculture
Panelists:
Prabhu Pingali, Deputy Director, Agricultural Development, The Bill and Melinda Gates Foundation (by phone)
Richard Broglie, Director of Research Strategy, DuPont Agricultural Biotechnology
Michael Roberts, Assistant Professor, Department of Agricultural and Resource Economics, North Carolina State University
Richard Van Atta, Senior Research Analyst, Science and Technology Policy Institute

5:00 PM Discussion

5:30 PM Poster Session

Presented by **AAAS FIRE** (Federal Innovation, Research, and Evaluation Affinity Group)

Mary Elizabeth Hughes , Science and Technology Policy Institute, Understanding High Risk, High-Reward Research Programs
Tiffany Sargent, National Science Foundation, Analytics for Managing Industrial and Government Portfolio Decisions
Amber Baum, National Science Foundation, The National Science Foundation's FY 2011 Performance Plan
Sapun Parekh, National Science Foundation, Flexible Portfolio Analysis of Fundamental Science and Engineering Research

Rebecca Rosen, National Institutes of Health, A Tool for Tracing, Understanding, and Visualizing NIH Contributions to Therapeutics Development

Kerry Hamilton, U.S. Environmental Protection Agency, Drinking Water Research Drivers and Future Directions

APRIL 19, 2011

8: 00 AM Registration

8:20 AM Welcome and Summary of First Day

Neal Lane, Co-Chair
Bronwyn Hall, Co-Chair

8: 30 AM Session VI: Impact of Research and Research Funding on the Labor Market and Career Development of STEM Professionals

How can better data and analysis on federal research spending be used to help the labor market function more efficiently? Is there a mismatch between the modes of funding graduate education and early career training and the labor market for STEM graduates? What kinds of data do we need to understand career preferences, career options, and career tracks especially in interdisciplinary fields?

Moderator: **Paula Stephan,** Professor of Economics, Georgia State University
Panelists:
Anthony Carnevale, Director, Georgetown University Center on Education and the Workforce
Henry Sauermann, Assistant Professor of Strategic Management, Georgia Institute of Technology
James Evans, Assistant Professor of Sociology, University of Chicago

9: 30 AM Discussion

10:00 AM Session VII: Emerging Metrics and Models for Assessing Research Impacts

What will it take to construct a long-term, comprehensive, disaggregated data infrastructure? Which challenges need the most attention? How can new approaches such as the STAR Metrics be improved and broadened to encompass different research programs, projects, performers, and funding mechanisms? How can advances in data presentation and visualization help policymakers better understand and use the analysis?

Moderator: **David Goldston**, Director of Governmental Affairs, Natural Resources Defense Council
Panelists:
Julia Lane, Program Director, Science of Science and Innovation Policy Program, National Science Foundation
Stefano Bertuzzi, Health Science Policy Analyst, Office of the Director, National Institutes of Health
Ian Foster, Arthur Holly Compton Distinguished Service Professor, Department of Computer Science, and Chan Soon-Shiong Scholar, University of Chicago
Lynne Zucker, Professor of Sociology and Policy Studies, University of California, Los Angeles
Adam Jaffe, Dean of Arts and Sciences and Fred C. Hecht Professor in Economics, Brandeis University
John Stasko, Professor and Associate Chair, School of Interactive Computing, Georgia Institute of Technology

11: 30 AM Discussion

12:00 PM Keynote Address: Subra Suresh, Director, National Science Foundation

Introduced by: **Michael Turner,** Rauner Distinguished Service Professor and Director, Kavli Institute for Cosmological Physics, University of Chicago

12:30 PM Lunch Break

1: 00 PM Session VIII: Impacts of Research on Decision-Making and Public Behavior

What is known about the impact of research on legislative, regulatory, and judicial decision-making? What do we know about the pathways by which advances in research eventually come to influence public behavior? Are there ways to enhance the effectiveness of these linkages?

Moderator: **Eric Ward**, President, The Two Blades Foundation
Panelists:
Kai Lee, Program Officer, Conservation and Science Program, David and Lucile Packard Foundation
Will Friedman, President, Public Agenda
Garry Neil, Corporate Vice President, Johnson and Johnson

2:30 PM Discussion

3:00 PM Session IX: Roundup Panel—Pitfalls, Progress, and Opportunities

Co-Moderators: **Neal Lane** and **Bronwyn Hall**, Co-Chairs
Panelists:
Alfred Spector, Vice-President, Google, Inc.
Eric Ward, President, The Two Blades Foundation
Paula Stephan, Professor of Economics, Georgia State University
David Goldston, Director of Governmental Affairs, Natural Resources Defense Council
Michael Turner, Rauner Distinguished Service Professor and Director, Kavli Institute for Cosmological Physics, University of Chicago

4:00 PM Adjourn

APPENDIX B

BIOGRAPHICAL INFORMATION

Speakers

NEAL LANE (Co-Chair) is the Malcolm Gillis University Professor at Rice University in Houston, Texas. He also holds appointments as senior fellow of the James A. Baker III Institute for Public Policy, where he is engaged in matters of science and technology policy, and in the Department of Physics and Astronomy. Lane served in the federal government during the Clinton administration as assistant to the president for science and technology and director of the White House Office of Science and Technology Policy (OSTP) from August 1998 to January 2001, and as director of the National Science Foundation (NSF) and member (ex officio) of the National Science Board from October 1993 to August 1998. Before becoming the NSF director, Lane was provost and professor of physics at Rice University, a position he had held since 1986. He first came to Rice in 1966, when he joined the Department of Physics as an assistant professor. In 1972, he became professor of physics and space physics and astronomy. Lane has received numerous prizes and awards, including the AAAS Philip Hauge Abelson Award, AAAS William D. Carey Award, American Society of Mechanical Engineers President's Award, American Chemical Society Public Service Award, American Astronomical Society/American Mathematical Society/American Physical Society Public Service Award, NASA Distinguished Service Award, Council of Science Societies Presidents Support of Science Award, Distinguished Alumni Award of the University of Oklahoma, the National Academy of Sciences Public Welfare Medal, the American Institute of Physics K.T. Compton Medal for Leadership in Physics and the Association of Rice Alumni Gold Medal for service to Rice University. Lane earned his B.S., M.S., and Ph.D. (1964) degrees in physics from the University of Oklahoma.

BRONWYN HALL (Co-Chair) is Professor in the Graduate School at the University of California at Berkeley and Professor of Economics of

Technology and Innovation at the University of Maastricht, Netherlands. She is a Research Associate of the National Bureau of Economic Research and the Institute for Fiscal Studies, London. She is also the founder and partner of TSP International, an econometric software firm. She received a B.A. in physics from Wellesley College in 1966 and a Ph.D. in economics from Stanford University in 1988. Professor Hall has published articles on the economics and econometrics of technical change, comparative analysis of the U.S. and European patent systems, the use of patent citation data for the valuation of intangible (knowledge) assets, comparative firm-level investment and innovation studies (the G-7 economies), measuring the returns to R and D and innovation at the firm level, analysis of technology policies such as R and D subsidies and tax incentives, and of recent changes in patenting behavior in the semiconductor and computer industries. She has also made substantial contributions to applied economic research via the creation of software for econometric estimation and of firm-level datasets for the study of innovation, including the widely used NBER dataset for U.S. patents. She is a member of the U.S. Federal Economic Statistics Advisory Committee, and the Research Advisory Councils of the Deutsche Bundesbank, Innovation Research Centre (University of Cambridge and Imperial College) and Solvay Business School (Brussels). She is also a past member of the Expert Group on Knowledge for Growth at the European Commission, and the Science, Technology, and Economic Policy (STEP) Board of the National Research Council.

STEFANO BERTUZZI is a Health Science Policy Analyst at the National Institutes of Health, Office of the Director. Bertuzzi is responsible for the NIH Return on Investment Program, in the Office of Science Policy, Office of the NIH Director, U.S. Department of Health and Human Services. In this position, Bertuzzi advises the NIH Director on a wide range of health science policy matters related to the impact of biomedical research on knowledge generation, health, wealth, and national competitiveness. Bertuzzi is the NIH lead for the STAR Metrics Project, which under the auspices of the White House Office of Science Technology and Policy aims at developing a novel infrastructure to capture the impact of federal R and D investments. Bertuzzi received his Ph.D. in Molecular Biotechnology at the Catholic University of Milan, Italy, and after postdoctoral training in the Laboratory of Molecular Neurobiology at the Salk Institute in San Diego, CA., became a tenured Associate Professor at the Dulbecco Telethon Institute in Milan, Italy.

RICHARD BROGLIE is Director of Research Strategy at DuPont Agricultural Biotechnology. He has a long history of research management in DuPont/Pioneer including trait discovery programs in the areas of improved soybean and canola oils and disease resistance in corn, soybean, wheat and rice. Currently he is responsible for agricultural biotechnology research programs in India, China, and Brazil as well as for the establishment of strategic public-private sector partnerships in these regions. Broglie received his Ph.D. in Microbiology from Rutgers University and served as both Postdoctoral Fellow and Assistant Professor at The Rockefeller University before joining DuPont in 1985.

ANTHONY CARNEVALE is the Director of the Georgetown University Center on Education and the Workforce. Between 1996 and 2006, Carnevale served as Vice-President for Public Leadership at the Educational Testing Service (ETS). While at ETS, Carnevale was appointed by President George Bush to serve on the White House Commission on Technology and Adult Education. Before joining ETS, Carnevale was Director of Human Resource and Employment Studies at the Committee for Economic Development (CED). While at CED, Carnevale was appointed by President Clinton to Chair the National Commission on Employment Policy. Carnevale was the founder and President of the Institute for Workplace Learning (IWL) between 1983 and 1993. While at the IWL, Carnevale was appointed by President Reagan to chair the human resources subcommittee on the White House Commission on Productivity between 1982 and 1984. Earlier, he was a senior staff member in both houses of the U.S. Congress. In 1993, President Clinton appointed Carncvale as chairman of the National Commission for Employment Policy. Carnevale received his B.A. from Colby College and his Ph.D. in public finance economics from the Maxwell School at Syracuse University.

PAUL CITRON is retired Vice President of Technology Policy and Academic Relations at Medtronic, Inc. Citron joined Medtronic in 1972 and worked in various positions until he retired in December 2003—Vice President of Science and Technology (1988-2002), Vice President, Ventures Technology (1985-1988), Vice President, Applied Concepts Research (1982-1985), Director, Applied Concepts Research (1979-1982), Design and Staff Engineer, Project and Program Manager (1972-1979). Citron was elected to the National Academy of Engineering in 2003, was elected Founding Fellow of the American Institute of Medical and Biological Engineering (AIMBE) in January 1993, has twice won the American College of Cardiology Governor's Award for Excellence and, in 1980, was inducted as a Fellow of the Medtronic Bakken Society. He was voted IEEE Young Electrical Engineer of the Year in 1979. He has authored many publications and holds several medical device pacing-related patents. In 1980 he was presented with Medtronic's "Invention of Distinction" award for his role as the co-inventor of the tined pacing lead. Citron received a B.S. in electrical engineering from Drexel University in 1969 and an M.S. in electrical engineering from the University of Minnesota in 1972.

CAROL CORRADO is senior advisor and research director in economics at The Conference Board. In addition, Corrado is a senior fellow of the Georgetown University Center for Business and Public Policy, and a member of the executive committee of the National Bureau of Economic Research's (NBER) Conference on Research on Income and Wealth. Corrado has authored key papers on the macroeconomic analysis of intangible investment and capital, including one that won the International Association of Research on Income and Wealth's 2010 Kendrick Prize ("Intangible Capital and U.S. Economic Growth") and one that appears in *Measuring Capital in the New Economy* (University of Chicago Press, 2005), a volume she co-edited. Previously, she was chief of the industrial output section at the Federal Reserve Board. Corrado received the American Statistical Association's prestigious Julius Shiskin Award for Economic Statistics in 2003 in recognition of her leadership in these areas and received a Special Achievement Award from the Board of Governors of the Federal Reserve System in 1998. She holds a Ph.D. in economics from the University of Pennsylvania and a B.S. in management science from Carnegie-Mellon University.

JAMES EVANS is Assistant Professor of Sociology and Fellow at the Computation Institute at the University of Chicago. Before coming to Chicago, he received his doctorate in sociology from Stanford University, served as a research associate in the Negotiation, Organizations, and Markets group at Harvard Business School, started a private high school in Utah focused on project-based arts education, and completed a B. A. in Anthropology from Brigham Young University. His current work explores how social and technical institutions shape knowledge—science, scholarship, law, news, religion—and how these understandings reshape the social and technical world.

IRWIN FELLER is senior visiting scientist at the American Association for the Advancement of Science and professor emeritus of economics at the Pennsylvania State University, where he has been on the faculty since 1963. Feller's long-time research interests include the economics of academic research, the university's role in technology-based economic development, and the evaluation of federal and state technology programs. He is the author of *Universities and State Governments: A Study in Policy Analysis* (Praeger Publishers, 1986) and many refereed journal articles. He has been a consultant to the President's Office of Science and Technology Policy, National Aeronautics and Space Administration, the Carnegie Commission on Science, Technology, and Government, the Ford Foundation, National Science Foundation, National Institute of Standards and Technology, COSMOS Corporation, SRI International, U.S. General Accounting Office, and the U.S. Departments of Education and Energy, among others.

IAN FOSTER is Arthur Holly Compton Distinguished Service Professor, Department of Computer Science, and Chan Soon-Shiong Scholor at University of Chicago. He is the Associate Division Director for Mathematics and Computer Science at Argonne National Laboratory and oversees the Distributed Systems Laboratory, which operates at both the University of Chicago and at Argonne National Laboratory. Foster's honors include the Lovelace Medal of the British Computer Society, the Gordon Bell Prize for high-performance supercomputing and an honorary doctorate from the Mexican Center for Research and Advanced Studies of the National Polytechnic Institute.

RICHARD FREEMAN holds the Herbert Ascherman Chair in Economics at Harvard University. He is currently serving as faculty co-director of the Labor and Worklife Program at the Harvard Law School. He directs the National Bureau of Economic Research / Sloan Science Engineering Workforce Projects, and is Senior Research Fellow in Labour Markets at the London School of Economics' Centre for Economic Performance. Freeman is a Fellow of the American Academy of Arts and Science. Freeman received the Mincer Lifetime Achievement Prize from the Society of Labor Economics in 2006. In 2007 he was awarded the IZA Prize in Labor Economics. In 2011 he was appointed Frances Perkins Fellow of the American Academy of Political and Social Science.

WILL FRIEDMAN joined Public Agenda in 1994, became associate director of research in 1996, and was the founding director of its public engagement department in 1997. In January 2011, he became president of Public Agenda. Friedman has overseen Public Agenda's expanding stream of work aimed at helping communities and states build capacity to tackle tough issues in more deliberative and collaborative ways. In 2007, he established Public Agenda's Center for Advances in Public Engagement (CAPE), which conducts action research to assess impacts and improve practice. He is also the co-editor, with Public Agenda chairman and co-founder Daniel Yankelovich, of the book, *Toward Wiser Public Judgment,* published in February 2011 by Vanderbilt University Press. Previously, Friedman was senior vice president for policy studies at the Work in America Institute, where he directed research and special projects on workplace issues. He was also an adjunct lecturer in political science at Lehman College, a research fellow at the Samuels Center for State and Local Politics, and a practitioner in the field of counseling psychology. He holds a Ph.D. in political science with specializations in political psychology and American politics.

DAVID GOLDSTON is Director of Government Affairs at the Natural Resources Defense Council in Washington, DC. Previously, Goldston served for six years as Chief of Staff of the House Committee on Science under Chairman Sherwood Boehlert of New York (2001-2006). Prior to becoming Chief of Staff, Goldston was Boehlert's legislative director during the years when Boehlert led a coalition of moderate Republicans that was pivotal in blocking environmental rollbacks. In that role, Goldston played a part in debates on a wide range of environmental issues, including clean air, forestry and endangered species. Goldston retired from the Congressional staff at the end of 2006 and has taught at Princeton and Harvard. He was also a monthly columnist on science policy issues for the journal *Nature.* Goldston graduated magna cum laude with a B.A. in American history from Cornell University in 1978. He completed the course work for a Ph.D. in American history at the University of Pennsylvania in 1993.

LAURA GUAY is Vice President of Research at the Elizabeth Glaser Pediatric AIDS Foundation. She is also a research professor at the George Washington University (GWU) School of Public Health and Health Services. She received her M.D. from GWU in 1985, and went on to a pediatrics residency at Rainbow Babies and Children's Hospital and Case Western Reserve University (CWRU) in Cleveland, Ohio. Guay was a visiting lecturer at Makerere University in Kampala, Uganda from 1988 to 1991, and then returned to CWRU to complete her fellowship in pediatric infectious diseases. She then spent seven more years in Uganda, where she worked on the landmark HIVNET 012 trial, which determined the effectiveness of single-dose nevirapine in preventing mother-to-child transmission of HIV. Prior to joining GWU,

Guay was a member of the faculty at the Johns Hopkins University School of Medicine. Most recently, her research has focused on reducing the rate of HIV transmission in breast-feeding infants and on the testing of an HIV vaccine in infants.

RUSH HOLT has represented central New Jersey in Congress since 1999. He earned his B.A. in Physics from Carleton College in Minnesota and completed his M.S. and Ph.D. at New York University. He has held positions as a teacher, Congressional Science Fellow, and arms control expert at the U.S. State Department where he monitored the nuclear programs of countries such as Iraq, Iran, North Korea, and the former Soviet Union. From 1989 until he ran for congress in 1998, Holt was Assistant Director of the Princeton Plasma Physics Laboratory, the largest research facility of Princeton University and the largest center for research in alternative energy in New Jersey. He has conducted extensive research on alternative energy and has his own patent for a solar energy device. In Congress Holt serves on the Committee on Education and the Workforce and the Committee on Natural Resources, where he serves as the ranking member on the Subcommittee on Energy and Mineral Resources. From 2007 to 2010, Holt was the Chairman of the Select Intelligence Oversight Panel.

ADAM JAFFE, the Fred C. Hecht Professor in Economics, has served as dean of the College of Arts and Sciences at Brandeis University since 2003. He has also held the position of chair of the economics department at Brandeis. Prior to joining the university in 1993, Jaffe was an assistant and associate professor at Harvard University and a senior staff economist at the President's Council of Economic Advisers. Jaffe's research focuses on the economics of innovation. His book *Innovation and Its Discontents: How Our Broken Patent System is Endangering Innovation and Progress, and What to Do About It*, co-authored with Josh Lerner was released in paperback in 2006. Jaffe earned a Ph.D. in economics at Harvard and an S.M. in technology and policy and an S.B. in chemistry from the Massachusetts Institute of Technology.

JULIA LANE is the Program Director of the Science of Science and Innovation Policy program at the National Science Foundation. Her previous jobs included Senior Vice President and Director, Economics Department at NORC/University of Chicago, Director of the Employment Dynamics Program at the Urban Institute, Senior Research Fellow at the U.S. Census Bureau and Assistant, Associate and Full Professor at American University. She became an American Statistical

Association Fellow in 2009. She is one of the founders of the LEHD program at the Census Bureau, which is the first large scale linked employer-employee dataset in the United States. A native of England who grew up in New Zealand, Julia has worked in Australia, Germany, Malaysia, Madagascar, Mexico, Morocco, Namibia, Sweden, and Tunisia. Her undergraduate degree was in Economics with a minor in Japanese from Massey University in New Zealand; her M.A. in Statistics and Ph.D. in Economics are from the University of Missouri in Columbia.

KAI LEE joined the David and Lucile Packard Foundation in 2007 as program officer with the Conservation and Science Program, where he is responsible for the science subprogram. Before joining the Foundation, Kai taught at Williams College from 1991 through 2007, and he is now the Rosenburg Professor of Environmental Studies, *emeritus*. He directed the Center for Environmental Studies at Williams from 1991–1998 and 2001–2002. Lee also taught from 1973 to 1991 at the University of Washington in Seattle. He holds a Ph.D. in physics from Princeton University and an A.B., *magna cum laude*, in physics, from Columbia University. He is the author of *Compass and Gyroscope* (1993). He is a member of the National Academies Roundtable on Science and Technology for a Sustainability Transition, and served most recently as vice-chair of the National Academies panel that wrote *Informing Decisions in a Changing Climate* (2009). Earlier, Lee had been a White House Fellow and represented the state of Washington as a member of the Northwest Power Planning Council. He was appointed in 2009 to the Science Advisory Board of the U.S. Environmental Protection Agency.

GARRY NEIL is the Corporate Vice President of Johnson and Johnson where he has held a number of senior positions within J and J, most recently Group President, Johnson and Johnson Pharmaceutical Research and Development . Under his leadership a number of important new medicines for the treatment of cancer, anemia, infections, central nervous system and psychiatric disorders, pain, and genitourinary and gastrointestinal diseases, gained initial or new and/or expanded indication approvals. Before joining J and JPRD, Neil held senior-level positions with Astra Merck Inc., Astra Pharmaceuticals, Astra Zeneca and Merck KGaA. He has also held a number of academic posts at a number of academic institutes including the Ludwig Institute for Cancer Research, the University of Toronto, the University of Iowa College of Medicine and the University of Pennsylvania (adjunct). He holds a Bachelor of Science degree from the University of Saskatchewan and a medical degree from the University of Saskatchewan College of Medicine and completed his postdoctoral clinical training in internal medicine and gastroenterology at the University of Toronto.

PRABHU PINGALI is the Deputy Director of Agricultural Development at the Bill and Melinda Gates Foundation. Formerly, he served as Director of the Agricultural and Development Economics Division of the Food and Agriculture Organization (FAO) of the United Nations. Pingali was elected to the U.S. National Academy of Sciences as a Foreign Associate in May 2007, and he was elected Fellow of the American Agricultural Economics Association in 2006. Pingali was the President of the International Association of Agricultural Economists (IAAE) from 2003-06. Pingali has over twenty five years of experience in assessing the extent and impact of technical change in agriculture in developing countries, including Asia, Africa and Latin America. From 1996-2002 he was Director of the Economics Program at Centro Internacional de Mejoramiento de Maíz y Trigo, Mexico. Prior to joining CIMMYT, from 1987 to 1996, he worked as an Agricultural Economist at the International Rice Research Institute at Los Baños, Philippines. Prior to that, he worked from 1982-1987 as an economist at the World Bank's Agriculture and Rural Development Department. He has received several international awards for his work, including two from the American Agricultural Economics Association: Quality of Research Discovery Award in 1988 and Outstanding Journal Article of the Year (Honorable Mention) in 1995. An Indian national, he earned a Ph.D. in Economics from North Carolina State University in 1982.

MICHAEL ROACH is Assistant Professor of Strategy and Entrepreneurship, Kenan-Flagler Business School at the University of North Carolina. Roach examines the sources and mechanisms by which firms utilize extramural knowledge in their innovative actives. In particular, his current research investigates how firms use university research in R and D activities and the subsequent impact of these knowledge flows on innovative performance. He also investigates how firms manage and protect intellectual capital, particularly through the strategic use and enforcement of patents. He teaches courses in entrepreneurial strategy, technological innovation and the management of intellectual capital. Roach was an entrepreneur before he became a professor. While in high school he co-founded a software start-up that specialized in the development of interactive educational programs for corporate executives and health care professionals. He also developed applications for handheld devices, including a system to aid primary health care workers in the diagnosis of communicable diseases. He received his Ph.D. in strategy from Duke University's Fuqua School of Business and his B.B.A. in decision sciences from Georgia State University's J. Mack Robinson College of Business.

MICHAEL ROBERTS is Assistant Professor of Agricultural and Resource Economics at North Carolina State University. Before joining the faculty at NCSU, Roberts worked for USDA's Economic Research Service. His research focuses on the intersection of agricultural and environmental economics. He has published papers on the effects of U.S. agricultural policies on production, land use, and the size of farms. Since leaving USDA, Roberts' research has focused increasingly on the potential effects of climate change on production of staple food grains and how biofuel growth has contributed to rising world food prices and food price variability. Roberts is also doing research on the design of procurement auctions, with an eye toward finding simple and cost-effective ways to buy environmental services like carbon sequestration from farmers and landowners.

BHAVEN SAMPAT is an Assistant Professor in the Department of Health Policy and Management at Columbia's Mailman School of Public Health. He also holds a courtesy affiliation with Columbia's School of International and Public Affairs (SIPA). An economist by training, Sampat is interested in issues at the intersection of health policy and innovation policy. His current projects examine the impacts of new global patent laws on innovation and access to medicines in developing countries, the political economy of the National Institutes of Health, the roles of the public and private sectors in pharmaceutical innovation, and institutional aspects of patent systems. Sampat has also written extensively on the effects of university patenting and entrepreneurship on academic medicine, and is actively involved in policy debates related to these issues. Sampat was previously an Assistant Professor at the School of Public Policy at Georgia Tech, where he won the "Faculty Member of the Year" teaching award in 2001-2002 and in 2002-2003. From 2003 to 2005 he was a Robert Wood Johnson Foundation Scholar in Health Policy Research at the University of Michigan. He is recipient of a Robert Wood Johnson Foundation "Investigator Award" to study how the NIH allocates its funds across disease areas.

MARCIO DE MIRANDA SANTOS is Executive Director of the Centre for Strategic Management and Studies in Science, Technology and Innovation and Chair of the Board of Trustees of the Center of

Reference on Environmental Information in Brazil. He received his M.Sc in Genetics and Plant Breeding and Ph.D. in Biochemical Genetics. He is also a former Visiting Scholar at Harvard University (1995-1997), where he studied the impacts of intellectual property rights regimes on the access and ownership of plant genetic resources utilized in food production and in other agriculture production systems. His major former professional appointments include: Director General, National Center for Genetic Resources and Biotechnology (1991-1995); Head, Brazilian Corporation for Agricultural Research (Embrapa) Department for Research and Development (1997-1999); Member and Chair of the International Plant Genetic Resources Institute Board of Trustees (1995-2002); Acting Director of Embrapa (1994-1995); and Professor of Evolutionary Biology, Catholic University of Brasilia (2000 to 2003). Santos was recently appointed as a member of the Consultative Group for International Agricultural Research Independent Scientific and Partnership Council.

DANIEL SAREWITZ is Professor of Science and Society at Arizona State University. Sarewitz's work focuses on understanding the connections between scientific research and social benefit, and on developing methods and policies to strengthen such connections. His most recent book is *Living with the Genie: Essays on Technology and the Quest for Human Mastery* (co-edited with Alan Lightman and Christina Desser; Island Press, 2003). He is also the co-editor of *Prediction: Science, Decision-Making, and the Future of Nature* (Island Press, 2000) and the author of *Frontiers of Illusion: Science, Technology, and the Politics of Progress* (Temple University Press, 1996). Prior to taking up his current position as director of the Center for Science, Policy, and Outcomes, he was the director of the Geological Society of America's Institute for Environmental Education. From 1989-1993 he worked on Capitol Hill, first as a Congressional Science Fellow, and then as science consultant to the House of Representatives Committee on Science, Space, and Technology, where he was also principal speech writer for Committee Chairman George E. Brown, Jr. Before moving into the policy arena he was a research associate in the Department of Geological Sciences at Cornell University, with field areas in the Philippines, Argentina, and Tajikistan. He received his Ph.D. in geological sciences from Cornell University in 1986.

HENRY SAUERMANN is Assistant Professor of Strategic Management at the Georgia Institute of Technology and holds a Ph.D. in Business Administration from Duke University. Dr. Sauermann's work examines the role of individuals' motives and incentives in shaping innovative activities and performance in organizations. One stream of his research examines the nature of scientists' pecuniary and nonpecuniary motives. This research also compares individuals' motives across organizational contexts and relates them to outcomes such as innovative performance in firms or patenting in academia. Another line of his work focuses on the goals and career choices of junior scientists and on the functioning of scientific labor markets.

BRIAN SLOAN is a senior policy analyst at the Research and Innovation Directorate General of the European Commission. A statistician by training, he started his career at the Commission in 1987 in Eurostat, the statistical office of the European Union. Since 1992 he has worked in the Commission department dealing with research policy and funding, where he has specialized in ex-post and ex-ante evaluation, and in the analysis and development of science and technology indicators.

ALFRED SPECTOR is Vice President for Research and Special Initiatives at Google, and is responsible for the research across Google and also a growing collection projects of strategic value to the company but somewhat outside the mainstream of current products. Previously, Spector was Vice President of Strategy and Technology IBM's Software Business, and prior to that, he was Vice President of Services and Software Research across IBM. He was also founder and CEO of Transarc Corporation, a pioneer in distributed transaction processing and wide area file systems, and was an Associate Professor of Computer Science at Carnegie Mellon University, specializing in highly reliable, highly scalable distributed computing. Spector received his Ph.D. in Computer Science from Stanford and his A.B. in Applied Mathematics from Harvard. He is a member of the National Academy of Engineering, a Fellow of the IEEE and ACM, and the recipient of the 2001 IEEE Computer Society's Tsutomu Kanai Award for work in scalable architectures and distributed systems.

JOHN STASKO joined the faculty at Georgia Tech in 1989, and is presently the Associate Chair of the School of Interactive Computing and Director of the Information Interfaces Research Group in the College of Computing. His primary research area is human-computer interaction, with a focus on information visualization and visual analytics. Stasko is also a faculty investigator in the Department of Homeland Security's VACCINE Center of Excellence focusing on developing visual analytics technologies and solutions for grand challenge problems in homeland security, and in the NSF FODAVA Center exploring the foundations of data analysis and visual analytics. He received the B.S. degree in Mathematics at Bucknell University in Lewisburg, Pennsylvania (1983) and Sc.M. and Ph.D. degrees in Computer Science at Brown University in Providence, Rhode Island (1985 and 1989).

PAULA STEPHAN is a Professor of Economics, Andrew Young School of Policy Studies, at Georgia State University and served as the founding associate dean of the school from 1996-2001. Her research interests focus on the careers of scientists and engineers and the process by which knowledge moves across institutional boundaries in the economy. Stephan's research has been supported by the Alfred P. Sloan Foundation, the Andrew Mellon Foundation, the Exxon Education Foundation, the National Science Foundation, the North Atlantic Treaty Organization and the U.S. Department of Labor. She has served on several National Research Council committees, is a regular participant in the National Bureau of Economics Research's meetings in Higher Education, and is a participant in the Science and Engineering Workforce Project based at the National Bureau of Economic Research. She currently is serving a three-year term as a member of the Advisory Board for the Social, Behavioral and Economic Sciences at the National Science Foundation. Dr. Stephan graduated from Grinnell College (Phi Beta Kappa) with a B.A. in Economics and earned both her M.A. and Ph.D. in Economics from the University of Michigan. Stephan coauthored with Sharon Levin *Striking the Mother Lode in Science*, published by Oxford University Press, 1992. Dr. Stephan has lectured extensively in Europe. She was a visiting scholar at the Wissenschaftszentrum Berlin für Sozialforschung, Germany, intermittently during the period 1992-1995.

SUBRA SURESH is the 13th director of the National Science Foundation (NSF). Prior to his confirmation as NSF director, Suresh served as Dean of the Engineering School and Vannevar Bush Professor of Engineering at the Massachusetts Institute of Technology (MIT). He joined MIT's faculty ranks in 1993 as the R.P. Simmons Professor of Materials Science and Engineering. During his more than 30 years as a practicing engineer, he held joint faculty positions in four departments at MIT as well as appointments at the University of California at Berkeley, Lawrence Berkeley National Laboratory and Brown University. Suresh has received many awards for his innovative research and commitment to improving engineering education around the world. Suresh is a co-inventor in more than 18 U.S. and international patent applications. He is author or co-author of several books that are widely used in materials science and engineering, including *Fatigue of Materials* and *Thin Film Materials*. He has consulted with more than 20 international corporations and research laboratories and served as a member of several international advisory panels and non-profit groups. Suresh has been elected to the U.S. National Academy of Engineering, American Academy of Arts and Sciences, Spanish Royal Academy of Sciences, German National Academy of Sciences, Academy of Sciences of the Developing World, Indian National Academy of Engineering and Indian Academy of Sciences. He earned his bachelor's degree from the Indian Institute of Technology in Madras in 1977, his master's from Iowa State University in 1979, and his doctorate from MIT in 1981.

MICHAEL TURNER is Bruce V. and Diana M. Rauner Distinguished Service Professor and Chair of the Department of Astronomy and Astrophysics at the University of Chicago. He also holds appointments in the Department of Physics and Enrico Fermi Institute at Chicago and is member of the scientific staff at the Fermi National Accelerator Laboratory. Turner received his B.S. in Physics from the California Institute of Technology (1971) and his Ph.D. in Physics from Stanford University (1978). His association with the University of Chicago began in 1978 as an Enrico Fermi Fellow and in 1980 he joined the faculty. Since 1979 Turner has been involved in the Aspen Center for Physics and served as its President from 1989 to 1993. From 2003 to 2005, Turner served as the Assistant Director of the National Science Foundation's Directorate for Mathematical and Physical Sciences. Turner is a Fellow of the American Physical Society and of the American Academy of Arts and Sciences and is a member of the National Academy of Sciences. He serves a member of the National Academies' Committee on Science, Engineering, and Public Policy. Turner's research interests are in theoretical astrophysics, cosmology, and elementary particle physics. He has made important contributions to inflationary Universe theory and understanding of dark matter.

RICHARD VAN ATTA is Senior Research Analyst at the Science and Technology Policy Institute (STPI). He came to STPI from the research staff of IDA's Studies and Analyses Center. His recent work has focused on advanced manufacturing issues and policies for effectively developing emerging technologies. Before joining IDA, he taught courses in national security and policy analysis at the American University's School of International Service, followed by several years of private consulting. From 1993 to 1998, Van Atta worked at the Department of Defense, first as Special Assistant for Dual Use Technology Policy and then as Assistant Deputy Under Secretary for Dual Use and Commercial Programs. Van Atta holds a BA in political science from the University of California, Santa Barbara, and a Ph.D. in political science from Indiana University.

IAN VINEY is Head of Strategic Evaluation at the Medical Research Council Head Office, United Kingdom. He gained a Ph.D. in genetics from Cambridge University in 1995. After a postdoctoral project at Imperial College he joined the MRC Head Office. Between 1998 and 2003, Viney worked as Research Programme Manager for genetics, and then Head of the Molecular and Cellular Medicine Board group. From 2003 to 2007 he was head of the MRC's Administrative Centre in London, working closely with MRC funded researchers at institutions across London on finance, personnel and strategic matters.

ERIC WARD is President of Two Blades Foundation, and was most recently CEO of Cropsolution, Inc., a crop protection chemical discovery company. Prior to that, he was Co-President of Novartis (now Syngenta) Agribusiness Biotechnology Research, where he was responsible for a staff of 270, including researchers and all administrative functions, including finance, patents, business development, public affairs, human resources, and facilities. Simultaneously, he was head of target discovery for Novartis Crop Protection AG, where he implemented a fully integrated agricultural chemical lead discovery program based on proprietary molecular targets. This program relied on extensive interactions with biotech firms and academic labs. Prior to that, he was a Research Director for the Novartis herbicide business unit, during which time his team invented Acuron™ herbicide tolerance technology, developed corn and sugar beet varieties engineered with the Acuron™ gene, and built the patent strategy to protect the technology. In 1994-5, he worked in Basel, Switzerland as a project leader for Ciba Crop Protection in the Weed Control business unit. He received his Ph.D. in plant biology from Washington University in St. Louis in 1988, where he was a graduate fellow of the National Science Foundation. He received his B.S in biology magna cum laude from Duke University in 1982.

BRUCE WEINBERG received his Ph.D. from the University of Chicago in 1996 before joining the faculty at the Ohio State University, where he is now Professor of Economics and Public Administration. He has held visiting positions at the Hoover Institution at Stanford, the National Bureau of Economic Research (NBER), the Kennedy School of

Government at Harvard; and the Federal Reserve Bank of Cleveland. He is a Research Associate at the NBER and a Research Fellow at the Institute for Labor (IZA), Bonn. He is an associate editor of the *New Palgrave Dictionary of Economics* and *Regional Science and Urban Economics* and currently serves as Director of Undergraduate Studies in Economics at the Ohio State University. His research has been supported by the Federal Reserve, the National Institutes of Health, the National Science Foundation, and the Templeton Foundation.

CATHERINE WOTEKI is the Under Secretary for Agriculture for Research, Education and Economics at the U.S. Department of Agriculture (USDA). Woteki served as Global Director of Scientific Affairs for Mars, Incorporated, where she manages the company's scientific policy and research on matters of health, nutrition, and food safety. From 2002-2005, she was Dean of Agriculture and Professor of Human Nutrition at Iowa State University. Woteki served as the first Under Secretary for Food Safety at USDA from 1997-2001, where she oversaw U.S. Government food safety policy development and USDA's continuity of operations planning. Woteki also served as the Deputy Under Secretary for Research, Education and Economics at USDA in 1996. Prior to going to USDA, Dr. Woteki served in the White House Office of Science and Technology Policy as Deputy Associate Director for Science from 1994-1996. Woteki has also held positions in the National Center for Health Statistics of the U.S. Department of Health and Human Services (1983-1990), the Human Nutrition Information Service at USDA (1981-1983), and as Director of the Food and Nutrition Board at the Institute of Medicine of the National Academy of Sciences (1990-1993). In 1999, Woteki was elected to the Institute of Medicine, where she chaired the Food and Nutrition Board (2003-2005). She received her M.S. and Ph.D. in Human Nutrition from Virginia Polytechnic Institute and State University (1974), and a B.S. in Chemistry from Mary Washington College (1969).

LYNNE ZUCKER is a Professor of Sociology and Policy Studies, and Director of the Center for International Science, Technology and Cultural Policy in the School of Public Policy and Social Research at the University of California, Los Angeles. Concurrently, she holds appointments as Research Associate with the National Bureau of Economic Research, and was previously a consulting sociologist with the American Institute of Physics. Zucker is the author of four books and monographs as well as numerous journal and other articles on organizational theory, analysis, and evaluation, institutional structure and process, trust production, civil service, government spending and services, unionization, science and its commercialization, and permanently failing organizations. Zucker received her A.B. with Distinction in Sociology and Psychology from Wells College in 1966. She received her M.A. in 1969 and Ph.D. in 1974 from the Sociology Department of Stanford University.

Staff

STEPHEN MERRILL (Project Director) has been executive director of the National Academies' Board on Science, Technology, and Economic Policy (STEP) since its formation in 1991. Dr. Merrill has directed many STEP projects and publications, including *Investing for Productivity and Prosperity* (1994); *Improving America's Schools* (1995); *Industrial Research and Innovation Indicators* (1997); *U.S. Industry in 2000: Studies in Competitive Performance and Securing America's Industrial Strength* (1999); *Trends in Federal Support of Research and Graduate Education* (2001), *A Patent System for the 21st Century* (2004), *Innovation Inducement Prizes* (2007), *Innovation in Global Industries* (2008), and *Managing University Intellectual Property in the Public Interest* (2010). For his work on patent reform he was recognized as one of the 50 leading world intellectual property experts by *Managing Intellectual Property* magazine and awarded the National Academies' 2005 Distinguished Service Award. Dr. Merrill's association with the National Academies began in 1985, when he was principal consultant on the report, *Balancing the National Interest: National Security Export Controls and Global Economic Competition.* In 1987 he was appointed to direct the National Academies' first government and congressional liaison office. Previously, Dr. Merrill was a Fellow in International Business at the Center for Strategic and International Studies (CSIS), where he specialized in technology trade issues. For seven years until 1981, he served on various congressional staffs, the last four years on that of the Senate Commerce, Science, and Transportation Committee. Dr. Merrill holds degrees in political science from Columbia (B.A., summa cum laude), Oxford (M. Phil.), and Yale (M.A. and Ph.D.) Universities. In 1992 he attended the Senior Managers in Government Program of the John F. Kennedy School of Government at Harvard University. From 1989 to 1996 he was an adjunct professor of international affairs at Georgetown University.

KEVIN FINNERAN is director of the Committee on Science, Engineering, and Public Policy, a joint unit of the National Academy of

Sciences, National Academy of Engineering, and the Institute of Medicine. He has been editor-in-chief of *Issues in Science and Technology* since 1991. Earlier he was Washington editor of *High Technology* magazine, a correspondent for the *London Financial Times* energy newsletters, and a consultant on science and technology policy. His clients included the National Science Foundation, the Office of Technology Assessment, the U.S. Agency for International Development, and the Environmental Protection Agency. Before launching his career in science and technology policy, he taught literature and film studies at Rutgers University. He is a fellow of the American Association for the Advancement of Science, the author of *The Federal Role in Research and Development* (1985), and a contributing author to *Future R and D Environments: A Report to the National Institute of Standards and Technology* (2002).

GURUPRASAD MADHAVAN is a program officer for the Committee on Science, Engineering, and Public Policy and the Board on Population Health and Public Health Practice at the National Academies. He has worked on such National Academies' publications as *Rising Above the Gathering Storm, Revisited: Now Approaching Category 5*; *The Reference Manual on Scientific Evidence (Third Edition)*; *Managing University Intellectual Property in the Public Interest*; and *Direct-to-Consumer Genetic Testing*. Madhavan received his B.E. (honors with distinction) in instrumentation and control engineering from the University of Madras, and his M.S. in biomedical engineering from State University of New York (SUNY) at Stony Brook. Following his medical device industry experience as a research scientist at AFx, Inc. and Guidant Corporation in California, Madhavan received an M.B.A., and a Ph.D. in biomedical engineering from SUNY Binghamton. Among other awards and honors, Madhavan was selected as one among 14 people as the "New Faces of Engineering" in the *USA Today* in 2009. He serves on the administrative council of the International Federation for Medical and Biological Engineering. Madhavan is co-editor of *Career Development in Bioengineering and Biotechnology* (Springer), *Pathological Altruism* (Oxford University Press), and *Practicing Sustainability* (Springer).

STEVE OLSON has been a consultant writer for the National Academies, the White House Office of Science and Technology Policy, the President's Council of Advisors on Science and Technology, the Howard Hughes Medical Institute, the National Institutes of Health, the Institute for Genomic Research, and many other organizations. From

1989 through 1992 he served as Special Assistant for Communications in the White House Office of Science and Technology Policy. Olson is the author of *Mapping Human History: Genes, Race, and Our Common Origins* (Houghton Mifflin), which was one of five finalists for the 2002 nonfiction National Book Award and received the Science-in-Society Award from the National Association of Science Writers. His book, *Count Down: Six Kids Vie for Glory at the World's Toughest Math Competition* (Houghton Mifflin), was named a best science book of 2004 by *Discover* magazine. His most recent book, cowritten with Greg Graffin, is *Anarchy Evolution* (HarperCollins). He has written several other books, including *Evolution in Hawaii* and *Biotechnology: An Industry Comes of Age*. He earned a bachelor's degree in physics from Yale University.

APPENDIX C

THE PROMISES AND LIMITATIONS OF PERFORMANCE MEASURES

Irwin Feller
Senior Visiting Scientist, American Association for the Advancement of Science and Professor Emeritus, Economics, Pennsylvania State University

I often say that when you can measure what you are speaking about, and express it in numbers, you know something about it; but when you cannot measure it, when you cannot express it in numbers, your knowledge is of a meager and unsatisfactory kind; it may be the beginning of knowledge, but you have scarcely in your thoughts advanced to the state of Science, whatever the matter may be.

-Baron William Thomson Kelvin.

When you can measure it, when you can express it in numbers, your knowledge is still of a meager and unsatisfactory kind.

-Jacob Viner

INTRODUCTION

Performance measurement is a politically powerful but analytical diffuse concept. Its meanings and implementation can vary from forcing fundamental changes in the ways in which public sector organizations and assessed and thus public funds allocated, as evinced by recent state government initiatives across all levels of U.S. education, to constituting old wine in new bottles, especially to empirically oriented economists, program evaluators and those weaned in the days of program-planning-budgeting.

Addressing this analytical diffuseness, this paper assesses the promises and limitations of performance measures as means of measuring economic and other returns of the Federal government's investments in basic and applied research. Referring to promises and limitations in the same sentence implies differences in perspectives and assessments about the relevance, reliability, validity, transparency, and suitability of performance measures to guide decision making. These differences exist. A stylized dichotomization is as follows:

- endorsement of requirements for, belief in, scholarly search supportive of, and opportunistic provision of performance measures that respond or cater to executive and legislative branch expectations or hopes that such measures will facilitate evidence-based decision-making;

- research and experientially based assessment that even when well done and used by adepts, performance measures at best provide limited guidance for future expenditure decisions and at worst are rife with potential for incorrect, faddish, chimerical, and counterproductive decisions.

The tensions created by these differences are best captured by the observation of Grover Cleveland, 22d and 24th President of the United States: "It's a condition we confront — not a theory." The condition is the set of Congressional and Executive requirements upon Federal agencies to specify performance goals and to provide evidence, preferably in quantitative form, that advances towards these goals have been made. The set includes by now familiar legislation such as the Government Performance and Results (GPRA) Act of 1993, the Government Performance and Results Modernization Act of 2010, and requirements of the 2009 American Recovery and Reinvestment Act's (ARRA) that Federal agencies provide evidence that their expenditures under the Act have stimulated job creation. It also includes comparable Executive branch directives. These include the Bush II Administration's articulation in 2002 of R and D Investment Criteria , subsequent implementation of these criteria by the Office of Management and Budget (OMB) via its Performance Assessment Rating Tool (PART) procedures, and the Obama Administration's 2009 OMB memorandum on Science and Technology Priorities for the FY 2011 Budget, that states that "Agencies should develop outcome-oriented goals for their science and technology activities…", and "… develop science of science policy" tools that can improve management of their research and development portfolios and better assess the impact of their science and technology investments."To these formal requirements may be added recent and likely increasing demands by congressional authorization and appropriations committees that agencies produce quantitative evidence that their activities have produced results, or impacts.

Theory here stands for a more complex, bifurcated situation, creating what Manski has termed dueling certitudes: internally consistent lines of policy analysis that lead to sharply contradictory predictions. (Manski, 2010). One theoretical branch is the currently dominant new public sector management paradigm branch. This paradigm emphasizes strategic planning, accountability, measurement, and transparency across all public sector functions, leading to, and requiring the use of evidence as the basis for informed decision making (OECD, 2005; Kettl, 1997).

The second branch is the accumulated and emerging theoretical and empirical body of knowledge on the dynamics of scientific inquiry and the processes and channels by which public sector support of research produces societal impacts. This body of knowledge performs a dual role. Its findings undergird many of the conceptualizations and expectations that policymakers have of the magnitude and characteristics of the returns to public investments in research and of the ways in which these returns can (or should) be measured. However, it is also a major source of the cautions, caveats, and concerns expressed by agency personnel, scientists, and large segments of the academic and science policy research communities that efforts to formally employ performance measures to measure public returns (of whatever form) to research and to then tie support for research to such measures are overly optimistic, if not chimerical, and rife with the potential for counterproductive and perverse consequences.

It is in the context of these differing perspectives that this paper is written. Its central thesis is that the promises and limitations of performance impact measures as forms of evidence relate to the decision-making context in which they are used. Context here means who is asking what type of question(s) with respect to what type of decision(s) and for what purpose(s). It also means the organizational characteristics of the Federal agency—can the activities of its operators be observed, and can the results of these activities be observed? (Wilson, 1989, pp. 158-171).

This emphasis on context produces a kaleidoscopic assessment, such that promises and limitations change shape and hues as the decision and organizational contexts shift. An emphasis on context also highlights the analytical and policy risks of assessing the promises and limitations of performance impact measures in terms of stylized characteristics. Performance measures for example can be used for several different purposes, such as monitoring, benchmarking, evaluation, foresight, and advocacy (making a case) (Gault, 2010). Consistent with the STEP-COSEPUP workshop's stated objective to provide expert guidance to Federal policymakers in the Executive and Legislative branches about what is known and what needs to be better known about how to assess economic and other returns to Federal investments in science and technology—the paper's focus is mainly on evaluation, although it segues at times into the other functions.

Approached in this way, performance is a noun, not an adjective. It also is a synonym for impact. This strict construction is made to separate the following analysis from the larger, often looser language associated with the topic in which performance is an adjective, as in the setting of strategic or annual (performance) goals called for by GPRA; as an indicator of current, changed or comparative (benchmarking) position, as employed for example in the National Science Foundation's biennial Science and Engineering Indicators reports; or as symptomatic measures of the health/vitality/position of facets of the U.S. science, technology and innovation enterprise, as represented for example in *Rising Above the Gathering Storm* (2007), where they are employed as evidence that things are amiss or deficient —a performance gap—in the state of the world.

The paper proceeds in a sequential, if accelerated manner. Section II contains a brief literature review and an outline of the paper's bounded scope. Section III presents a general discussion of the promises and limitations of performance measures to assess the impacts of Federal investments in research. Section IV illustrates the specific forms of the promises and limitations of performance measures in the context of what it terms the "big" and "small" questions in contemporary U.S. science policy. Section V offers a personal, "bottom line" perspective on what all this means.

Analytical Framework and Scope

The paper's analytical framework and empirical findings derive mainly from economics, although its coverage of performance measures is broader than economic statistics and its treatment of impact assessment is based mainly on precepts of evaluation design. The choice of framework accords with the workshop's objective, which is suffused with connotations of efficiency in resource allocation, or more colloquially, seeking the highest possible returns on the public's (taxpayer's) money. Adding to the appropriateness and relevance of the chosen approach is that many of the arguments on behalf of Federal investments in research, both basic and applied, draw upon economic theories and findings. As Godin has noted, "We owe most of the quantitative analysis of S and T to economists" (Godin, 2005, p. 3).

An immediate consequence of treating the workshop's objective in this manner is that a goodly number of relevant and important subjects, policy issues, and analytical frameworks are touched upon only briefly, while others are ignored completely. Thus, only passing attention is taken of the historical, institutional and political influences that in fact have shaped and continue to shape the setting of U.S. national science priorities and Federal R and D budgets, whether viewed in terms of allocations by broad objectives, agencies, fields of science, or modes of support. Moreover, interpreting the workshop's objective as a search for measures related to allocative efficiency obviously sidesteps topics and rich streams of research related to political influences on national research priorities (e.g., Hedge and Mowery, 2010) or which generate earmarks, set asides, and sheltered capacity building competitions that palpably diverge from efficiency objectives (e.g., Savage, 1999; Payne, 2006). Likewise omitted are consideration of the normative goals underlying Federal support of research and the distributive effects or societal impacts that flow from it (Bozeman and Sarewitz, 2011).

Another consequence is that the paper is primarily about performance measurement as a generic approach rather than about the reliability and validity of specific measures. Where reference is made to specific measures, it is to illustrate larger themes. In fact, there is no shortage of "metrics", in GPRA-speak- to measure the outputs and outcomes of Federal investments in research. Geisler (2000; pps. 254-255) offers a well presented catalogue of 37 "core" metrics. These are organized in terms of immediate outputs (e.g., number of publications in refereed journals; number of patents); intermediate outputs (e.g., number of improved or new products produced; cost reductions from new and improved products/processes); pre-ultimate outputs (e.g., savings, cost reductions, and income generated by improved health, productivity, safety, and mobility of the workshop at sectoral and national levels); and ultimate outputs (e.g., improved GDP/capital; improved level of overall satisfaction and happiness of population.) The list is readily expanded to include combinations of single indicators, new data sets that permit disaggregation of existing measures, and new and improved versions of mainstream measures–the rapid and seemingly accelerating move from publication counts to citation measures to impact factors to h-indices and beyond being one such example.

Also in abundance are various scorecards or rankings based on assemblages and weightings of any number of performance measures related to scientific advance, technological advance, competitiveness, innovativeness, university performance, STEM-based educational proficiency and the like that have been used to position US performance within international hierarchies or norms. Indicator construction for science and technology has become a profession in its own stead, with regular international conferences—The European Network of Indicators Designers will hold its 2011 Science and Technology Indicators Conference in Rome, Italy, in September, 2011— and a well recognized set of journals in which new work is published.

Plentiful too and continuously being updated are compendia and manuals covering international best practice on how to evaluate public sector R and D programs. These works cover a wide range of performance impact measures and methodologies, including benefit-cost analysis, patent analysis, network analysis, bibliometrics, historical tracings, innovation and on the outputs produced by several different Federal agencies—health, energy, agriculture, environmental protection, international competitiveness, employment. (For recent overviews, see Wagner and Flanagan, 1995; Ruegg and Feller, 2003; Godin, 2005, chpt. 15; Kanninen and Lemola, 2006; Grant, et.al, 2009; Foray, 2009; Gault, 2010; Link and Scott, 2011).

Finally, in setting expectations for the workshop, it is perhaps helpful to note that the topics and issues to be discussed are not new ones. Rather, they form the substance of at least 60 years of theoretical, empirical and interpretative work, producing what by now must be a five foot high stack of reports and workshop proceedings, including a sizeable number originating under National Academies' auspices. The recurrent themes addressed in this previous work, evident since the program-planning-budgeting initiatives of the 1960s and continuing on through its several variants, are a search for decision algorithms that will lead to the improvement in government budgeting and operations and a search for criteria for setting priorities for science (Shils, 1969). Noting these antecedents is not intended to diminish the importance of current activities (nor, for that matter, of this paper). Instead, it is to suggest the complexities of the issues under consideration and as a reminder of the richness and contemporary relevance of much that has been written before.

Performance Impact Measures

Differences in assessments about the potential positive and negative features of requiring strategic plans and performance measures into how Federal agencies set research priorities and assessed performance were visible at the time of GPRA's enactment. They continue to this day.[1]

In 1994, almost immediately after GPRA's passage, I organized a session at the American Association for the Advancement of Science's (AAAS) Colloquium on Science and Technology Policy on the applicability of GPRA to budgeting for science and technology. Taking a "neutral" stand on the subject, I invited, among other panelists, Robert Behn, a leading scholar of and advocate for the new public management paradigm subsumed within GPRA and like requirements, and Paul David, a leading researcher in the economics of science and technology.

The title of Behn's talk captured its essence: "Here Comes Performance Assessment-And it Might Even be Good for You." (Behn,

---

[1] A natural experiment occurring on February 15-16, 2011 highlights the continuing character of these differing   perspectives. OSTP's release on February 10, 2011 of its R and D Dashboard, that contains data about NIH and NSF R and D awards to research institutions and "links those inputs to outputs—specifically publications, patent applications, and patents produced by researchers funded by those investments"– produced an immediate flurry of comments and exchanges on SciSIP's list server. Most of this exchange contained the   point-counterpoint themes in the Behn-David exchange cited above, as well as those recounted in this paper. Among these were: how were outcomes defined? could they be measured? is there reasonable consensus on what they are? One rejoinder to these comments raised in response to specific reservations about the meaningfulness of patent data was that when Congress asks what are we getting from these billions spent on R and D, it is helpful to have patent number to point to as one outcome of the nation's investment.

1994). Among the several benefits (or promises) cited by Behn were the following:

Having objectives ("knowing where you want to go") is helpful;Objectives provide useful baseline for assessing each of 4 modalities of accountability–finance, equity, use of power and performance.

Well defined objectives and documentation of results facilitate communication with funders, performers, users, and others.

For his part, David outlined what he termed "very serious problems…with outcome goal setting for federal programs in general and for research in particular" (David, 1994, p. 294). David's central argument was that an "outcome reporting may have a perverse effect of distorting the perception of the system of science and technology and its relationship to economic growth" (ibid, p. 297). He further observed that " Agencies should define appropriate output and outcome measures for all R and D programs, but agencies should not expect fundamental basic research to be able to identify outcomes and measure performance in the same way that applied research or development are able to."
What follows is essentially an expanded exposition of these two perspectives, presented first as promises and then as limitations.

Promises

- Performance measurement is a (necessary) means towards implementing (and enforcing) the audit precepts – especially those linked to accountability and transparency–contained within GPRA and like requirements.

- Performance measures can assist agencies make improved, evidence-based decisions both for purposes of program design and operations (formative evaluations) and longer term assessments of allocative and distributive impacts (summative evaluations). In these ways, performance measures assist agencies in formulating more clearly defined, realistic, and relevant strategic objectives and in better adjusting ongoing program operations to program objectives.

- Well defined, readily measured, and easily communicated performance measures aids both funders and performers to communicate the accomplishments and contributions of the public investments to larger constituencies, thereby maintaining and strengthening the basis of long term public support of these investments.

- The search for measures that accurately depict what an agency/program has accomplished may serve as a focusing device, guiding attention to the shortcomings of existing data sets and thus to investments in obtaining improved data.

- Performance measurement focuses attention on the end objectives of public policy, on what has happened or happening outside the black box, rather than on the churning of processes and relationships inside the black box. This interior churning produces intermediate outputs and outcomes (e.g., papers, patents) that may be valued by performers (or their institutions, stakeholders, or local representatives), but these outputs and outcomes do not necessarily connect in a timely, effective, or efficient manner to the goals that legitimize and galvanize public support.

- Requiring agencies to set forth explicit performance research goals that can be vetted for their societal importance and to then document that their activities produced results commensurate with these goals rather than some diminished or alternative set of outputs and outcomes is a safeguard against complacency on the part of funders and performers that what might have been true, or worked in the past, is not necessarily the case today, or tomorrow. Jones, for example, has recently noted, "Given that science is change, one may generally imagine that the institutions that are efficient in supporting science at one point in time may be less appropriate at a later point in time and that science policy, like science itself, must evolve and continually be retuned" (Jones, 2010, p. 3). Measurement of impacts is one means of systematically attending to the consequences of this evolution.

- Performance measurement is a potential prophylactic against the episodic cold fusion-type viruses that have beset the formulation of U.S. science policy. As illustrated by the continuing debates set off by Birch's claims on the disproportionate role of small firms as sources of job creation (cf. Haltiwanger, J., R. Jarmin, and J. Miranda (2010)) or the challenge posed to the reflexive proposition that the single investigator mode of support is the single best way to foster creative science by Borner, et.al. findings that "Teams increasingly dominate solo scientists in the product of high-impact, highly cited science; (Borner, et. al. 2010, p. 1), U.S. science and innovation policy contains several examples of Will Roger's observation that, "It isn't what we don't know that gives us trouble, it's what we know that ain't so."

- Presented as a method of assessing returns to Federal investments in research, performance measurement provides policymakers and performers with an expanded, more flexible and adaptable set of measures than implied by rate of return or equivalent benefit-cost calculations. Criticism of what is seen as undue reliance on these latter approaches is longstanding; they are based in part on technical matters, especially in the monetization of non-market outputs, but also on the distance between the form that an agency's research output may take and the form needed for this output to have market or other societal impacts.

The largest promise of performance measurement, though, likely arises not from recitation of the maxims of the new public management but from the intellectual ferment now underway in developing new and improved data on the internal processes of scientific and technological research, the interrelationships of variables within the black box, and improved methods for assembling, distilling and presenting data. Much of this ferment, of course, relates to Dr. Marburger's call for a new science of science policy, the activities of the National Science and Technology Committee's (NSTC) Committee on Science, and the research currently being supported by the National Science Foundation's Science of Science and Innovation Policy program (SciSIP). No attempt is made here to present a full précis of the work underway (Lane, 2010). Having been a co-organizer, along with Al Teich, of two AAAS workshops at which SciSIP grantees presented their preliminary findings and interacted with Federal agency personnel, however, it is a professional pleasure to predict that a substantial replenishment and modernization of the intellectual capital underlying existing Federal research policies and investments can be expected.

To illustrate though the nature of recent advances, I cite two developments non-randomly selected to reflect the focus of my own research interests. They are the NSF's Business R and D and Innovation Survey (BRDIS), itself in part redesigned in response to the 2005 NRC study, *Measuring Research and Development Expenditures in the U.S. Economy*, and advances in the visualization of the (bibliometric) interconnections of disciplines. The NRC report articulated longstanding concerns that NSF's existing survey of industrial R and D needed methodological upgrading, lagged behind the structure of the U.S.

economy in not adequately covering the growth of the service sector or the internationalization of sources and performers of R and D, and did not adequately connect R and D expenditures with downstream "impact" measures, such as innovations. The result has been a major revision of these surveys, undertaken by NSF's Science Resources Statistics Division.

Early findings from the new BRDIS survey on the sources and characteristics of industrial innovation fill a long recognized data gap in our understanding of relationships between and among several variables, including private and public R and D expenditures, firm size and industrial structure, human capital formation and mobility, and managerial strategies. (Boroush, 2010). Combined with pending findings from a number of ongoing SciSIP projects and juxtaposed to and compared with data available from ongoing international surveys, these newly available data hold promise of simultaneously providing policymakers with a finer grained assessment of the comparative and competitive position of the technological performance of the U.S. economy and researchers and evaluators finer grained data to assess the impacts of selected science and technology program and test existing and emerging theories.

Science is a set of interconnected, increasingly converging disciplines, so run the claims of many scientists (Sharp, et. al, 2011). But precisely in what ways and with what force do these interconnections flow? Does each field influence all other fields and with equal force, or are there discernible, predictable differences in patterns of connection and influence? Prospectively, being able to answer these questions would provide policymakers with evidence about relative priorities in funding fields of science, presumably giving highest priority to those that served as hubs from which other fields drew intellectual energy. Recent research in data visualization, illustrated by Boyack, Klavans, and Borner's *Mapping the Backbone of Science* (2005), combines bibliometric techniques, network theories, and data visualization techniques to offer increasingly accessible "maps" of the structure of the natural and social sciences, thereby providing one type of answer to these questions.

Limitations

The above noted emphasis on context surfaces immediately in considering the limitations of performance measurement. Perhaps the most obvious and important difference in the use of such measures in between ex ante and ex post decision making settings. Fundamental differences exist in the theoretical, analytical and empirical knowledge bases for using performance measures to determine whether past investments have produced the research expected of them and using such measures to decide upon the magnitude and direction of new funds.

If retrospective assessment was all that was implied by the call for performance measures of impacts, the task before this audience, and for Federal science agencies in satisfying new planning and reporting requirements, while challenging, especially in reconciling and distilling divergent, at times conflicting findings, as say in the cases of the Bayh-Dole Act (Larsen, 2010; NRC, 2010) or the SBIR program's generation of sustainable increases in employment, would at least be relatively straightforward. There is no shortage of well crafted assessments of past Federal investments in basic and applied research. Such work has been and continues to be a staple component of research on the economics of science and technology and of previous NRC reports over the past 50 years. A short list would include the rich empirical literatures on the returns to Federal investments in agricultural research (Evenson, Ruttan, and Waggoner, 1979; Heisey, et. al, 2010), biomedical research (Murphy and Topel, 2006; Stevens, et. al, 2011), energy efficiency research (Link, 2010), and applied technology programs, such as NIST's Advanced Technology Program (Ruegg and Feller, 2003). [2]

Manifestly though, more than an assessment of past investments as a possible guide to future decisions is intended in recent requirements and continuing calls for measures of performance impact for research. The central premise underlying the mantra of evidence-based decision making is that some combination of findings about the impacts from previous findings or findings from some form of in situ or heuristic experiment provides the best possible predictor of the expected impacts that will follow upon Federal research expenditures. This is a far different matter than assessing the impacts of past research expenditures.[3] The premise though must confront considerable research-

---

[2] An added value of calling attention to these retrospective studies is that they all entail studying the relationship between a cause and an effect: "between the activities involved in a public program and any outcome of that program..." (Mohr, 1995, p. 1). Without a theoretical foundation that specifies the set of dependent and independent variables to be, performance measures represent what Koopman's has termed, "measurement without theory." (1947, p. 161).

[3] The implicit assumption throughout this paper is that decisions follow or at least are influenced by, the evidence contained in performance measures. That

based agnosticism of many scholars of the extent to which findings based on past studies can be used to forecast the specific magnitude and characteristics of future Federal investments in research. As noted by Crespi and Guena, for example, "After more than 50 years scholarly work on the importance of academic research, there is still little systematic evidence on how such investments can lead to increase levels of scientific output, improved patenting and innovative output, better economic performance and, ultimately, to increase national wealth" (Crespi and Geuana, 2008, p. 555).

The primary limitation of using performance measures to shape future Federal investments in research flows from the well documented tale, widely recounted in both the scholarly and policy literatures, that the outcomes of scientific research are unpredictable as to when they will occur, who will be responsible for them, and even more so with respect to their end uses. This last influence appears to be of increasing importance in confounding projections of returns to future Federal research investments as 'users" become increasingly influential in transforming platform scientific findings or technological advances into their own new products and processes (von Hippel, 2005).

Additionally, again to restate familiar propositions, the impacts of basic and applied research occur only over extended periods of time, often extending beyond budgetary and planning horizons. They also frequently require further "investments"—downstream in terms of prototype development, manufacturability, and marketing–and upstream, in terms of related scientific discoveries or technological breakthroughs– before their impacts are felt. Many, if not most of these necessary complementary activities are outside the purview of the agencies funding the research.

To cite two examples from the literature on the economics and history of science and technology that express these propositions. First, Rosenberg: "From an economic point of view, perhaps the most striking peculiarity of knowledge production is that is not possible to establish

---

is, evidence of good/high performance leads to the continuation/expansion of a program; evidence of poor/low performance leads to termination/contraction. This is obviously a stylized proposition. Technically sophisticated assessments of the contributions of the ATP program did little to save it from the congressional budget axe, and in light of the current political environment one can anticipate that a similar fate awaits Federal research programs related to environmental protection and climate change, however well done and rich with societal impacts they may be.

the nature of its production function. We can never predict the output which will be generated by a given volume of inputs. By its very nature knowledge production deals with forays into the unknown. It involves the combination of resources to an exploratory process the outcome of which may be a large number of dead ends rather than the hoped-for-discovery of knowledge or techniques possessing profitable economic applications." (Rosenberg, 1972, p. 172). Second, Mowery and Rosenberg: "It is essential to emphasize the unexpected and unplanned, even it-or especially if-it renders serious quantification impossible. In fact, the difficulties in precisely identifying and measuring the benefits of basic research are hard to exaggerate" (Mowery and Rosenberg, 1989, p. 11).

These assessments are widely shared. To cite an earlier National Academies endeavor not unlike today's, "History...shows us how often basic research in science and engineering leads to outcomes that were unexpected or took many years or even decades to emerge...The measures of the practical outcomes of basic research usually must be retrospective and historical and...the unpredictable nature of practical outcomes is an inherent and unalterable feature of basic research" (National Academies, 1999). They are also found in Executive budget documents. For example, although suffused with an emphasis on quantitative performance measures, OMB's earlier articulation of R and D Investment Criteria expressed nuanced understanding of the uncertainties surrounding returns from Federal investments in basic research: "Agencies should define appropriate output and outcome measures for all R and D programs, but agencies should not expect fundamental basic research in the same way that applied research or development are able to do. Highlighting the results of basic research is important, but it should not come at the expense of risk-taking and innovation" (OMB, PART 2008, Appendix C, p. 76).

To briefly illustrate these propositions with some specifics, the above tale is well told by the historical linkages between and among the work of physicists, Pauli, Purcell, and Bloch in identifying and measuring nuclear magnetic resonance (NMR); the use of NMR by chemists to determine the structure of molecules; the sequential development by Varian of increasingly more user-friendly NMR machinery, and the subsequent, still contested priority race between supporters of Damadian and Lauterbur to apply NMR to medical imagining, along with a host of other advances in mathematics, computer science, and technologies, a number of which originated with firms such as EMI in the United Kingdom and GE in the U.S., leading to the now ubiquitous presence of MRI (Kelves, 1997; Roessner, et. al, 1997).

My personal favorite example of the meanderings of new scientific and technological knowledge into uses not anticipated by those funding or performing the underlying research is the answer I received from Penn State undergraduates enrolled in my course in science and technology in the pre-IPod, circa 2000 period, when asked to identify the most pressing national S and T policy issue. The overwhelming response was the then legal imbroglio relating to downloading Napster files. So much for DARPA and the search to link high-end, computing-intensive research.

The limitations of performance measures as forms of evidence to guide investments in research extend beyond this general case. There are other specific limitations that arise in or bear upon specific decisions in specific contexts. A partial listing of them is as follows:

- Performance measures for research may undervalue lack of performance, or failure. Science has been described as the only field where failure is to be expected. After all, it was Edison who when challenged with the number of failed experiments rejoined, "I have not failed. I've just found 10,000 ways that won't work." Indeed, given the well known skewness of the distribution of research findings, at least as measured by bibliometric data, recent agency initiatives to promote high-risk, high-reward research if interpreted formally implies increased frequency of projects that fail to achieve their stated objectives.

- There is an implied but at times illusive exactitude in first speaking about the promise/benefits of performance measurement and then moving to the selection and operationalization of specific performance measures. As illustrated by current national debates over the specification of performance measures in K-12 and higher education, the transition is seldom that simple. For example, is the performance of public colleges, and thus their state appropriations under a system of performance based budgeting, to be based on graduation rates, time to degree, mastery of general knowledge, mastery of specialized knowledge, or life time earnings, preferably within state borders? These measures reflect different concepts of performance, several of which point provide different sets of incentives for university administrators and faculty. Alternative articulations of performance likewise are subsumed within the global objectives set for Federal research programs–productivity increase, for example, is not synonymous with job creation.

- Leaving aside issues associated with data availability and quality, the casual linkages between program/agency objectives and the choice of measures to be used can be fuzzy. Empirically, reservations can be expressed about the metrics that agencies have to date employed and been accepted by OMB to document agency performance. My earlier brief of the performance metrics contained in OMB's PART review pointed to a diverse set of measures across agencies and programs. Some related to technical specifications (i.e., achieve a certain level of performance advance), some to economic gains (threshold and above benefit-cost ratios), some to societal impacts (e.g., reduction in traffic fatalities), and more. No clear analytical or empirical distinction though seemed to be made between what seemed in some cases to be final impacts and in others to be intermediate impacts. There is an admitted logic to this variability. Agencies differ not only in objectives, but in the technical ease with which it is possible to measure performance relative to these objectives. It is easier to measure rates of return to commodity-oriented agricultural research, where market data on inputs and outputs are readily available, than to investments in research on particle physics. But the hodgepodge of performance measures in use undercuts any attempt to systematically compare performance across agencies in formulating research budgets. In practice, the specification of the measures to be used, as well as the target to be reached, likely are the outcome of some form of negotiation and compromise between OMB and the agency, perhaps with some formal or informal understanding of what is acceptable to the relevant congressional committee. This is conjecture, though, awaiting confirmation. At present, the specification of performance measures across agencies and programs should be viewed as a new policy and research black box.

- In a circular process, unless a program's objective is defined in terms of a single performance measure, any single measure is at best a partial indicator of the objective being pursued. In most cases, single performance measures are only loosely connected to higher order performance objectives. Moreover, employment of single measures can produce findings that incorrectly suggest that the objectives for which research support is being provided are not being met, and thus be misleading guides to public policy. Hill's 1986 study, undertaken for the House Committee on Science and Technology, of the relationship between US Nobel Prize awards and aggregate U.S. performance objectives in economic growth and health highlights these risks (U.S. Congress, 1986; p.65). As illustrated in Figure C-1, Hill's study, undertaken at the height of angst about U.S. international economic competitiveness, shows a negative relationship in growth in gross domestic product and national Nobel Prize awards in physics and chemistry!

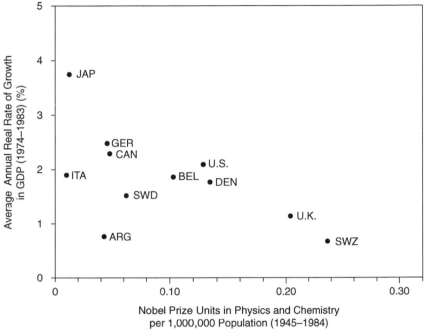

FIGURE C-1 Growth in Gross Domestic Product and Nobel Prize-winning in Physics and Chemistry.

Connecting U.S. performance in Nobel prizes to health outcomes produces only a slightly better "picture." The mapping of Nobel prizes in physiology and medicine for the period 1945-1984 with health statistics shows a slight positive relationship in reductions in infant mortality but no apparent association to gains in life expectancy. Compare this use of this single performance measure with Cutler and Kadiyala's estimate that an average 45-year old in 1994 had a life expectancy 4 ½ years longer than in 1950 because cardiovascular disease mortality had decreased. This finding leads them to the "unambiguous conclusion …that medical research on cardiovascular disease is clearly worth the cost" (2002, p.113). Cast in benefit-cost terms, this increase is estimated to yield a 4 to 1 return for medical treatment and a 30-to-1 return for research and dissemination costs related to behavioral change.

More generally, to the extent that single measures are used, they become the de facto criterion by which performance is judged, and in a system of performance-based budgeting the basis on which decisions about which future funding is made. Measures, though, are means to ends. As the recent if by now jaded maxim put it, if you can't measure it, you can't manage it. But measures also can shape the ends. What is measured is what is managed, or promised. The value of impacts not measured is thereby diminished, or ignored. More tellingly, what is measured is what is produced if measured performance is tied to resources or rewards of whatever form.

An obvious implication of the vignettes from health research is the risk of using any single performance measure to gauge the impacts of Federal investments in research. This proposition is stated so frequently and explicitly in contemporary exegesis on assessment of science programs (Schmoch, et. al, 2010) that it would not be worth mentioning, except that it is frequently ignored. Thus, the contributions of a university's research activities to national or state level economic growth are often reduced to counts of numbers of patents or licenses, or even worse to the ratio of patents to external R and D funding, while the job creation expectations associated with the research funding in ARRA have taken on a life of its own, creating an assessment cynosure about that performance measure to the exclusion of other, possibly broader and deeper impacts (Goldston, 2009).

- Most federal research programs, though, have multiple rather than single objectives. Multiplicity creates its own set of problems for use of performance measures on impacts. It increases the prospects for non-commensurable findings because of intractable technical issues of measuring things that are heterogeneous within a single comprehensive measure and because of the implicit normative weighting accorded different objectives (Gladwell, 2011). The presence of multiple objectives also increases the likelihood of variable performance (satisfactory, unsatisfactory; results not proven) among them, leaving open or requiring normative assessments about relative weights. The setting also is rife with the prospects of strategic retreats on objectives, so that performance becomes measured in terms of what an agency/program can produce, not what it was set up to produce. Moving to multiple measures, as in scorecards, also raises questions of possible co-linearity among seemingly independent measures, so that what seems to be the richness of the approach in effect reduces to variants of a single measure. Perhaps most importantly, the presence of multiple objectives for a program increases the likelihood of trade-offs among facets of performance, so that an increase in one agency/program objective can be achieved only at the expense of a decrease in another objective. A substrata tension, or inconsistency, thus seems to exist between the simultaneous pursuit of multi-item performance measurement scorecards and acceptance of the organizational mantra that one can't be all things to all people.

- Effectiveness and efficiency are different concepts. This difference is frequently overlooked in interpreting measures of the performance of Federal programs. It is one thing to say that a program has produced positive outcomes with respect to one or more of its several stated objectives; it is another to say that it is achieving these outcomes in the most efficient manner relative to the next best uses to which its program funds could have been or could be put. It is axiomatic that any large scale Federal program, research or otherwise, especially if longstanding, will have many accomplishments to report. A corollary is that the larger and longer standing the program, the larger the absolute number of outputs.

  The potential consequences for misinterpreting evidence and subsequent questionable decisions when effectiveness and efficiency are confounded takes on special importance in light of the recent OMB memorandum, "Increased Emphasis on Program Evaluations." The memorandum states that "Rigorous, independent program evaluations can be a key resource in determining whether government programs are achieving their intended outcomes as well as possible and at the lowest possible cost." It also notes that, "And Federal programs have rarely evaluated multiple approaches to the same problem with the goal of identifying which ones are most effective." Absent some form or control or comparison group or other explicit standard, performance measures provide little basis for determining a program's cost-effectiveness or efficiency.

- The informational content of performance measures may change over time. This may result from employing measures in different ways than they had formerly been used, especially if a different set of incentives is attached to them—faculty member's patenting of their research shifting over time in promotion and tenure reviews from a negative indicator of distraction from disinterested Bohr- cell research to a positive indicator of fulfillment of a university's "third mission" objectives in promotion and tenure packets. It may also change as a result of legislative or court decisions and/or firm strategies in the uses made of the measured activity- reported shifts, for example, in increased patent activities by firms and the trend towards using them as a source of revenue as well as a means of protecting intellectual property (Cohen, 2005). Freeman and Soete formalize and label this changing informational content as the science, technology, and innovation (STI) version of Goodhart's law: "once STI indicators are made targets for STI policy, such indicators lose most of the informational content that qualify them to play such a role" (2009, p. 583). Along similar lines, Moed, in his review of bibliometric indicators has noted the argument that indicators applied in research performance assessment should be modified every ten years or so, replacing indicators normally applied by new types (Moed, 2005, p. 320).

The Use of Performance Measures in Federal Research Policy Decisions

Context matters critically as one attempts to relate the above characteristics of performance measures to the type of decisions that U.S. policy makers are called upon to make with respect to Federal investments in research. The applicability and thus promise and limitations of performance measures, singly or collectively, can vary greatly from holding high potential for providing useful information on program and project level activities to low, problematic and potentially counterproductive for overarching decisions concerning levels and broad allocation patterns of Federal support.

Schematically and historically, U.S. science and technology policy has consisted of a continuing discourse between a stock of big questions and a comparable stock of big answers. This discourse underpins the continuity of the main features of U.S. policy, albeit with short-term economic and political perturbations. It also provides the intellectual and policy capital base for consideration of a continuing flow of smaller questions and smaller answers about specific science and innovation policy issues. These latter issues flare up to dominate near-term science policy forums, and then through some amalgam of a modicum of resolution, overtaking by the eruption of new policy agenda items, or by morphing into the big questions lose their immediate saliency, only to pop again anew, not infrequently with new terms being used to describe recurring questions.

The big three science policy questions, in the U.S. as well as elsewhere, are (1) the optimal size of the federal government's investments in science and technology programs; (2) the allocation of these investments among missions/agencies/and programs and thus fields of science; and (3) the selection of performers, funding mechanisms, and the criteria used to select projects and performers.[4] Permeating each of these questions is the question of "why," namely, the appropriate, effective and efficient role of the Federal government in supporting public investments in science and technology including nurturance of a STEM-qualified labor force. Allowing for variations in language and precipitant events, the questions are strikingly unchanged between the 1960s ferment on criteria for scientific choice and those posed by Dr. John Marburger's call for a new science of science policy. Only the first and second questions are treated here.

The big answers to these questions appear in the sizeable and ever more sophisticated theoretical, descriptive and empirical literature that has dealt with these topics since at least the 1960s. These answers form the basis for statement in the NSTC's 2008 report, *The Science of Science Policy: a Federal Research Roadmap* termed, a "...well developed body of social science knowledge that could be readily applied to the study of science and innovation."

The big answers, in summary form, relate to: (1) The contributions of productivity increase to increases in GDP/capita. This answer is the intellectual and policy legacy of a continuing stream of work from Solow-Abramovitz, through the growth accounting work and debates of economists such as Denison, Jorgensen, Baumol, more recently recast in terms of endogenous growth theory and spillover effects. Thus, the opening paragraph of the National Academies highly influential report, *Rising Above the Gathering Storm,* states that "Economic studies conducted even before the information-technology revolution have shown that as much as 85% of measured growth in U.S income per capital was due to technological change." (2) Market failure propositions associated with the work of Arrow and Nelson that competitive markets

---

[4] "The major issues in science policy are about allocating sufficient resources to science, to distribute them wisely between activities, to make sure that resources are used efficiently and contribute to social welfare" (Lundvall, B. and S. Borras, 2005, p. 605)

fail to supply the (Pareto-) optimal quantity of certain types of R and D[5] and (3) Mansfield-Griliches-type analytical frameworks, augmented increasingly with attention to knowledge spillovers, based on divergences between social and private rates of return to R and D, that have been used to justify Federal investments across a swathe of functional domains - health, agriculture, environmental protection.

Historically, these big answers have become the ideas of academic scribblers that influence those in power in power today, or least most of them. They have contributed to shaping a broad political consensus, ranging from President Obama to George Will, about the appropriateness, indeed necessity, of Federal support of basic research, at the same time leaving in dispute the legitimacy of Federal support for technologically oriented civilian-oriented research programs.

Because these answers are so much a part of contemporary discourse, it is easy today to lose sight of their transformative impacts. These answers invert most of core tenets of pre-1950 Federal science and technology policy in which support was provided for mission oriented, applied research but not for basic research (DuPree, 1957). Likewise, the current major role of U.S. universities as performers of Federally funded basic research–a role today much valued, extolled and defended by these institutions–had to overcome fears expressed by leaders of the National Academy of Sciences about "government interference with the autonomy of science…" (Geiger, 1986, p. 257).

But even as the broad policy propositions derived from the big answers remain essentially correct in shaping the overall contours of US science policy, they are seen as of limited value by decision makers because in their view the answers do not correspond to the form in which they confront decisions about how much to investment in research and how these investments should be allocated among national government objectives, fields of science, and agencies (Feller, 2009).

For example, in addressing the first overarching question of how much the Federal government should spend in the aggregate to R and D, abstracting here from thorny analytical, measurement, and institutional

---

[5] These propositions have turned out to be a double-edged theoretical sword for purposes of Federal research policy. They were initially advanced to justify public sector support for basic research but as interpreted in the 2002 OSTP-OMB R and D Investment Criteria and then implemented in the PART process, they have become the theoretical basis for excising several domestic technology development programs.

issues involved in linking R and D with technical change and/or productivity change, that the business sector funds two-thirds of US R and D, and that total Federal R and D expenditures are the sum of multiple House and Senate appropriations' bills, how does one move from the 85% share of growth in per capita income attributed to technological progress contained in *Rising Above the Gathering Storm* to findings such as Boskin and Lau's estimate that 58% of the economic growth of the United States between 1950-1998 was attributable to technical progress to determining the optimal R and D/GDP ratio? How would the optimal level of Federal expenditures on R and D change if new findings suggested that existing estimates overstated the contribution of technical progress by 20/30 percentage points, or conversely, understated this contribution by a like amount?

Using a different performance measuring stick, given candid assessments from European officials that the European Union's 2000 Lisbon Strategy's 3% goal was a political rather than an economic construct, what's the empirical basis of the Obama Administration's 3% goal? Achieving, or surpassing the goal would reverse declines in real terms in Federal support of R and D since 2004 (Clemins, 2010), and raise the ratio from its current estimated level of about 2.8%. Achieving this goal would also move the US closer to the top of all other OECD nations, even possibly overtaking Finland or Sweden (Boroush, 2010). But other than asserting that more is better than less, what other basis exists for determining whether 3% is too high, too low, or just right? Given all the above cited reservations about the complexity of linking public sector research expenditures to desired outcomes, how can performance measures be used to exist to judge the merit of recent proposals that the U.S. should be spending 6 %, not 3% of GDP on R and D (Zakaria, 2010)?

Similar questions arise when or if one attempts to start from treating science not as the handmaiden of economic growth but having its own internal dynamics. The cover page of the 125th anniversary issue of *Science*, 1 July 2005, is titled: 125 Questions: "What Don't We Know?" Assume, as seems appropriate, that these questions accurately represent the current frontiers of knowledge, that advances in many if not all directions at the frontier hold the promise of societal benefits, and that excellent proposals addressing each question are awaiting submission to the relevant Federal agency. What would be the total cost? What share of GDP or of the Federal budget would be required to fund these proposals once added to other national or mission agency R and D priorities? If not all these proposals could be funded, what means should be used to select from among them? What measures of performance/output/outcomes should be used to assess past performance in determining-out year investments or near-term R and D priorities. Exciting as it may be to envision the prospects of societal impacts flowing from frontier, high-risk, transformative risk, it serves only to bring one full circle back to the policymaker's priority setting and resource allocation questions noted above.

The same issues arise when trying to compute the proper level of support or estimate the returns to public investments for functional objectives, agencies, and fields of science. An impressive body of research, for example, exists on the contributions to the health status of the American population produced by Federal investments in biomedical research. It's an analytical and empirical stretch to say that this research provides evidence that can be used to determine whether current or proposed levels of appropriation for NIH are right, too little, or too high. No evident empirical basis existed for the doubling of NIH's budget over a 5-year time period, and the consequences now observed while unintended were not unpredictable (Freeman and van Reenan, 2008). At issue here is what Sarawitz had termed the myth of infinite benefits: "If more science and technology are important to the well-being of society, then the more science and technology society has, the better off it will be" (1996; p. 18). Indeed, arguably, if the budget decision had any lasting impacts, it was to elevate "balance" of funding across agencies as a resource allocation criteria and to set doubling as a formulaic target for other science oriented agencies.

Similar problems arise too in attempting to formulate analytically consistent criteria based on performance measures for allocating funds among fields of science and technology, — how much for chemistry?; physics?; economics?- especially as among national objectives and agencies, as well as within agencies. These are the perennial practical questions across the spectrum of Federal science policymakers, yet perhaps with the exception of basing program level allocations on estimated returns from impacts, as in the cases of agriculture (Ruttan, 19820 and health (Gross, Anderson, and Power, 1999) for which few good answers, or funding algorithms, exist. For example, a recent NRC panel tasked with just such an assignment concluded in its report, *A Strategy for Assessing Science*, "No theory exists that can reliability predict which research activities are most likely to lead to scientific advances or to societal benefits" (2007, p. 89).

One would like to do better than this. Here, if anywhere, is where performance measurement may have a role. The challenge at this point is not the absence of performance measures relating Federal investments in research to specific outputs or studies pointing to high social rates of return within functional areas but the sheer number of them and the variations in methodologies that produce them. The result is a portfolio of options about performance measures, each more precisely calibrated over time but still requiring the decision maker to set priorities among end objectives.

Thus, the Boyack, et. al, bibliometric study cited above highlights the "centrality" of biochemistry among published papers. Using this study and its implied emphasis on scientific impact as a basis for resource allocation decisions among scientific fields would presumably lead to increased relative support for biochemistry. If one instead turns to the Cohen-Nelson-Walsh survey-based study (2002) of the contributions of university and government laboratory research i.e., ("public") research to industrial innovation, which contains an implied policy emphasis on economic competitiveness, one finds both considerable variation across industries in the importance of public research and variations in which fields of public research are cited as making a contribution. An overall finding though is that, "As may be expected, more respondents consider research in the engineering fields to contribute importantly to their R and D than research in the basic sciences, except for chemistry" (2002, p. 10). The authors however mute this distinction of the relative contribution of fields of science with the caution that "the greater importance of more applied fields does not mean that basic science has little impact, but that its impact may be mediated through the more applied sciences or through the application of industrial technologists' and scientists' basic scientific training to the routine challenges of conducting R and D" (p. 21). But the upshot of the study still would seem to be the need for increased (relative) support of engineering related disciplines. Advocates for increased Federal research for computer science and engineering, for their part may turn to Jorgenson, Ho, and Samuels' recent estimates of the contribution of the computer equipment manufacturing industry to the growth in US productivity between 1960-2007 (Jorgenson, Ho, and Samuels, 2010).

An obvious conclusion, indeed the standard one in discussion of this issue, is that the interconnectedness of fields of science requires that each be supported. And this of course is how the present U.S. system functions. There are considerable differences, however, between funding each field according to its deeds and each according to its needs. Moreover, the interconnectedness argument applies to historical determinants and levels of support; it is of limited guidance in informing budget decision—show much more or less, given existing levels of support?

Little of this should be a surprise. The gap between estimates of returns to public investments in research and using these estimates to formulate budget allocations among missions, agencies, and disciplines was identified by Mansfield in the opening text of the social returns to R and D. Referring to the number of independent studies working from different models and different data bases that have pointed to very higher social rates of return, he noted, "But it is evident that these studies can provide very limited guidance to the Office of Management and Budget or to the Congress regarding many pressing issues. Because they are retrospective, they shed little light on current resource allocation decisions, since these decisions depend on the benefits and costs of proposed projects, not those completed in the past" (Mansfield, 1991, p. 26). The gap has yet to be closed.

Similar issues arise in using bibliometric data to allocate resources across fields. Over the last three decades, even as the U.S. position in the life sciences has remained strong, its world share of engineering papers has been cut almost in half, from 38 percent in 1981 to 21 percent in 2009, placing it below the share (33 percent) for the EU27. Similar declines in world share are noted for mathematics, physics, and chemistry (National Science Foundation, 2007; Adams and Pendlebury, 2010). One immediate, and simple interpretation of these data is that aggregate bibliometric performance is a function of resource allocation: a nation gets what it funds. But this formulation begs first the question if what it is producing is what it most needs, and then if what it is producing is being produced in the most efficient manner.

Conclusion

Having studied, written about, participated in, organized workshops on, and as an academic research administrator been affected by the use of performance measures, something more than an "on the one hand/on the other hand" balance sheet, a concluding section seems in order.

It's simpler to start with the limitations of performance measures for they are real. These include the attempt to reduce assessment of complex, diverse, and circuitously generated outcomes, themselves often dependent on the actions of agents outside the control of Federal agencies, to single or artificially aggregated measures; the substitution of bureaucratically and/or ideologically driven specification and utilization of selective measures for the independent judgment of experts; and the distortion of incentives for science managers and scientists that reduces the overall performance of public investments. To all these limitations must be added that to date there is little publically verifiable evidence outside the workings of OMB-agency negotiations that implementation of a system of performance measurement has appreciably improved decision making with respect to the magnitude or allocation of Federal research funds. When joined with reservations expressed by both scholars and practitioners about the impacts of the new public management paradigm, it produces assessments of the type, "Much of what has been devised in the name of accountability actually interferes with the responsibilities that individuals in organizations have to carry out work and to accomplish what they have been asked to do" (Radin, 2006, p.7; also Perrin, 1998; Feller, 2002; Weingert, 2005; Auranen and Niemien, 2010).

The promises, too, are likely to be real if and when they are realized. One takes here as a base the benefits contained in Behn's presentation and the section on promises above. Atop this base are to be added the revised and new, expanded, disaggregated, and manipulable data sets emerging both from recent Federal science of science policy initiatives and other ongoing research (Lane and Bertuzzi, 2011). Thus, Sumell, Stephan, and Adams' recent research on the locational decisions on new Ph.D.s working in industry accords with and provides an empirical base for the recent calls by the National Science Foundation's Advisory Committee for GPRA Performance Assessment 2008 to collect and provide data on the "development of people" as an impact of agency support.

A different category of benefits owing less to improved public sector management practices and more to the realities of science policy decision making needs to be added to this list. The very same arguments cited above that the links between initial Federal investments in research are too long term and circuitous to precisely specify in GPRA or OMB planning or budget formats serves to increase the value for intermediate measures. For policymakers operating in real time horizons, even extending beyond the next election cycle, performance measures of the type referred to above are likely as good as they are to get. We live in a second best world. Although it may be analytically and empirically correct to state say that none of the proximate intermediate output measures, patents or publications for example, are good predictors of the ultimate impacts that one is seeking–increased per capita income, improved health–some such measures are essential to informed decision making.

Adding impetus to this line of reasoning is that the environment in which U.S. science policy is made is a globally competitive one, which increases the risks of falling behind rivals. Akin to an arms race or advertising in imperfectly competitive markets, support of research is necessary to maintain market share, even if the information on which decisions are made is imperfect.

Finally, as an empirically oriented economist whose work at various times has involved generating original data series of patents and publications and use of a goodly portion of the performance measures and methodologies now in vogue in evaluations of Federal and State science and technology programs, there is a sense of déjà vu to much of the debate about the promises and limitations of performance measures of impacts. The temptation is to observe somewhat like Monsieur Jordain in Moliere's play, Le Bourgeois Gentilhomme, "Good heavens! For more than forty years I have been doing performance measurement without knowing it."

Performance measures viewed either or both as a method for explicating needs assessments or conducting impact assessments are basic, indispensible elements in policy making, program evaluation, and scholarly research. What are open to issue are:

- the specification of the appropriate measures for the decision(s) under review– a complex task involving technical, political, and normative considerations;

- the proper interpretation and incorporation of existing and newly developed data and measures used in retrospective assessments of performance into decisions relating to estimating the prospective returns from alternative future Federal investments in research– decisions made within a penumbra of scientific, technical, economic, and societal uncertainties that performance measures reduce but do not eliminate; and

- providing evidence that use of performance measures as forms of evidence in fact improves the efficiency or rate(s) of return to Federal investments in research.

Given the above recitation of promises and limitations, the optimal course of action seems to be what Feuer and Maranto have termed science advice as procedural rationality (2010). It is to (1) have policymakers employ performance impact measures that correspond to what is known or being learned about how public investments in basic and applied science relate to the attainment of given societal objectives; (2) have the body of existing and emerging knowledge of how Federal in basic and applied research impact on societal objectives connect to the form of the decisions that policymakers are called upon to make; and (3) use gaps that may exist between (1) and (2) to make explicit the nature of the limits to which theory-based/evidence-based knowledge can contribute to informed decision making (Aghion, David, and Foray, 2009). Viewed in terms of preventing worse case outcomes, the objective should be to avoid the pell-mell drive now in vogue in State governments towards formula shaped coupling of performance measures and budgets, a trend as applied to Federal funding of research that is fraught with the risks of spawning the limitations described above.

To the extent that the STEP-COSEPUP workshop contributes to producing this course of action, it will have made an important contribution to the formulation of US research policy.

REFERENCES

Adams, J., and Pendlebury, D. 2010. Global Research Report: United States (Thomas Reuters).

Aghion, P., David, P., and Foray, D. 2009. Can We Link Policy Practice with Research on 'STIG' Systems? Toward Connecting the Analysis of Science, Technology and Innovation Policy with Realistic Programs for Economic Development and Growth, *The New Economics of Technology Policy*, edited by D. Foray (Cheltenham, UK: Edward Elgar):46-71.

Auranen, O. and Nieminen, M. 2010. University Research Funding and Publication Performance-An International Comparison. *Research Policy*, 39:822-834.

Behn, R. 1994. *Here Comes Performance Assessment-and It Might Even be Good for You. AAAS Science and Technology Policy Yearbook-1994*, edited by A. Teich, S. Nelson, and C. McEnaney (Washington, DC: American Association for the Advancement of Science), 257-264.Borner, K., Contractor, N., Falk-Krzesinski, H. J., Fiore, S. M., Hall, K., L., Keyton, J., Spring, B., Stokols, D., Trochin, W., and Uzzi, B. 2010. A Multi-Systems Perspective for the Science of Team Science. *Science Translational Medicine*, 2:1-5.

Boroush, M. 2010. New NSF Estimates Indicate that U.S. R and D Spending Continued to Grow in 2008. *National Science Foundation Infobrief,* 10-32. Arlington, VA: National Science Foundation, January 2010.

2010. NSF Releases New Statistics on business Innovation, *National Science Foundation Info Brief,* 11-300, October 2010. Arlington, VA: National Science Foundation.

Boskin, M., and Lau, L. 2000. Generalized Solow-Neutral Technical Progress and Postwar Economic Growth NBER Working Paper 8023 Cambridge, MA: National Bureau of Economic Research.

Bozeman, B., and Sarewitz, D. 2011. Public Value Mapping and Science Policy Evaluation. *Minerva,* 49:1-23.

Clemins, P. 2010. Historic Trends in Federal R and D in Research and Development FY2011 Washington, DC: American Association for the Advancement of Science , 21-26.

Cohen, W. R. 2005. Patents and Appropriation: Concerns and Evidence. *Journal of Technology Transfer,* 30:57-71

Cohen, W., Nelson, R., and Walsh, J. 2002. Links and Impacts: The Influence of Public Research on Industrial R and D. *Management Science,* 48:1-23.

Crespi, G. , and Geuna, A. 2008. An Empirical Study of Scientific Production: A Cross Country Analysis, 1981-2002. *Research Policy,* 37: 565-579.

Cutler, D., and Kadiyala, S. 2003. The Return to Biomedical Research: Treatment and Behavioral Effects, in *Measuring the Gains from Medical Research*, edited by K. Murphy and R. Topel, Chicago, IL: University of Chicago Press, pp. 110-162.

David, P. 1994. Difficulties in Assessing the Performance of Research and Development Programs. *AAAS Science and Technology Policy Yearbook-1994*, op. cit., 293-301.

DuPree, A. H. 1957. *Science in the Federal Government.* New York: Harper Torchbooks.

Executive Office of the President, Office of Management and Budget, Science and Technology Priorities for the FY2012 Budget, M-10-30.

Evenson,R., Ruttan, V., and Waggoner, P. E. 1979. Economic Benefits from Research: An Example from Agriculture. *Science* 205: 1101-1107.

Feller, I. 2002. Performance Measurement Redux. *American Journal of Evaluation*, 23:435-452(2007).

Feller, I. Mapping the Frontiers of Evaluation of Public Sector R and D Programs. 2007. *Science and Public Policy*, 34:681-690.

Feller, I. 2009. A Policy-Shaped Research Agenda on the Economics of Science and Technology . *The New Economics of Technology Policy*, edited by D. Foray .Cheltenham, UK: Edward Elgar, 99-112.

Feller, I. and G. Gamota. 2007. Science Indicators as Reliable Evidence. *Minerva*, 45:17-30.

Feuer, M. and Maranto ,C. 2010. Science Advice as Procedural Rationality: Reflections on the National Research Council. *Minerva:* 48:259-275.

Freeman, C. and Soete, L. 2009. Developing Science, Technology and Innovation Indicators: What We Can Learn from the Past. *Research Policy*, 38:583-589.

Freeman, R. and van Reenan , J. 2008. Be Careful What You Wish For: A Cautionary Tale about Budget Doubling. *Issues in Science and Technology, Fall* .Washington, DC: National Academies Press.

Gault, F. 2010. *Innovation Strategies for a Global Economy.* Cheltenham, UK: Edward Elgar.

Geiger, R. 1986. *To Advance Knowledge.* Oxford, UK: Oxford University Press.

Geisler, E. 2000. *The Metrics of Science and Technology.* Westport, CT: Quorum Books.

Gladwell, M. 2011. The Order of Things. *New Yorker*, February 14, 2011: 68ff

Godin, B. 2005. *Measurement and Statistics on Science and Technology.* London, UK: Routledge.

Goldston, D. 2009. Mean What You Say. *Nature* 458, 563. Published online 1 April 2009.

Grant, J., Brutscher, P., Kirk, S., Butler, L., and Woodring, S. 2009. Capturing Research Impacts: A Review of International Practice, *Report to the Higher Education Funding Council for England*, DB-578-HEFCE.Cambridge, UK: RAND Europe.

Gross, C., Anderson, G., and Powe, N. 1999. The Relation between Funding by the National Institutes of Health and the Burden of Disease. *New England Journal of Medicine*, 340 (24):1881-1887

Haltiwanger, J., Jarmin, R., and Miranda, J. 2010. Who Creates Jobs? Small vs. Large vs. Young. *National Bureau of Economic Research Working Paper 16300.* Cambridge, MA: National Bureau of Economic Research.

Hedge, D. and Mowery, D. 2008. Politics and Funding in the U.S. Public Biomedical R and D System. *Science*, 322 (19):1797-1798

Heisey, P., King, K., Rubenstein, K., Bucks, D., and Welsh, R.2010. Assessing the Benefits of Public Research Within an Economic Framework: The Case of USDA's Agricultural Research Service. United States Department of Agriculture, Economic Research Service, *Economic Research Report Number 95.*

Jones, B. 2010. As Science Evolves, How Can Science Policy? *National Bureau of Economic Research Working Paper 16002.* Cambridge, MA: National Bureau of Economic Research.

Jorgenson, D., Ho, M. and Samuels, J. 2010. New Data on U.S. Productivity Growth by Industry. Paper presented at the World KLEMS Conference, Harvard University, August 19 - 20, 2010.

Kanninen, S., and Lemola, T. 2006. *Methods for Evaluating the Impact of Basic Research Funding.* Helsinki, Finland: Academy of Finland.

Kelves, B. 1997. *Naked to the Bone.* New Brunswick, NJ: Rutgers University Press.

Kettl, D. 1997. The Global Revolution in Public Management: Driving Themes, Missing Links. *Journal of Policy Analysis and Management,* 16:446-462.

Koopmans, T.J. 1947. Measurement without Theory. *Review of Economic Statistics,* 39:161-172

Lane, J. and Bertuzzi, S. 2011. Measuring the Results of Science Investments. *Science,* 331(6018):678-680.

Larsen, M. 2011. The Implications of Academic Enterprise for Public Science: An Overview of the Empirical Literature. *Research Policy,* 40:6-10.

Link, A. 2010. Retrospective Benefit-Cost Evaluation of U.S. DOE Vehicle Combustion Engine R and D Investments. Department of Economics Working Paper Series.

Link, A. and Scott, J. 2011 *Public Goods, Public Gains,* Oxford, UK: Oxford University Press.

Lundvall, B. and Borras, S. 2005. Science, Technology, and Innovation Policy. *The Oxford Handbook of Innovation*, edited by J. Fagerberg, D. Mowery, and R. Nelson. Oxford, UK: Oxford University Press. 599-631.

Mansfield, E. 1991. Social Returns from R and D: Findings, Methods and Limitations. *Research Technology Management,* 34:6

Manski, C. 2010. Policy Analysis with Incredible Certitude. NBER Working Paper Series #16207. Cambridge, MA: National Bureau of Economic Research.

Massachusetts Institute of Technology (2011) The Third Revolution: The Convergence of the Life Sciences, Physical Sciences, and Engineering. Letter to our Colleagues, January 2011.

Moed, H. 2005. *Citation Analysis in Research Evaluation*. Dordrecht, The Netherlands: Springer.

Mohr, L. 1995. *Impact Analysis for Program Evaluation*. 2d Edition Thousand Oaks, CA: SAGE.

Mowery, D. and Rosenberg, N. 1989. *Technology and the Pursuit of Economic Growth*. Cambridge, UK: Cambridge University Press.

Murphy, K. and Topel, R. 2006. The Value of Health and Longevity. *Journal of Political Economy*. 114:871-904.

National Academies of Sciences.1999. *Evaluating Federal Research Programs* .Washington, DC: National Academy Press.

National Academies of Sciences. 2007. *Rising Above the Gathering Storm*. Washington, DC: National Academies Press.

National Academies of Sciences. 2007. *A Strategy for Assessing Science*. Washington, DC: National Academies Press.

National Academies of Sciences. 2010. *Managing University Intellectual Property in the Public Interest*. Washington, DC: National Academies Press.

National Science Foundation 2007. Changing U.S. Output of Scientific Articles: 1988-2003 *–Special Report* . Arlington, VA: National Science Foundation.

Office of Management and Budget .2008. *Program Assessment Rating Tool Guidance,* No. 2007-02

Organisation for Economic Cooperation and Development. 2005. *Modernising Government*. Paris, FR: Organisation for Economic Cooperation and Development.

OECD Science, Technology and Industry Outlook. 2010. Paris, FR: Organiszation for Economic Cooperation and Development.

Payne, A. 2006. Earmarks and EPSCoR. *Shaping Science and Technology Policy*, edited by D. Guston and D. Sarewitz. University of Wisconsin Press. 149-172.

Perrin, B. 1998. Effective Use and Misuse of Performance Measurement. *American Journal of Evaluation,* 19: 367-379.

Radin, B. 2006. Challenging the Performance Movement. Washington, DC: Georgetown University Press.

Roessner, D.,Bozeman, B. , Feller, I., Hill, C., and Newman, N. 1997. *The Role of NSF's Support of Engineering in Enabling Technological Innovation, Report to the National Science Foundation* Arlington, VA: SRI International.

Rosenberg, N. 1972. *Technology and American Economic Growth*. New York, NY: Harper Torchbooks.

Rosenberg, N. 1982. Learning by Using, in *Inside the Black Box* Cambridge, UK: Cambridge University Press. 120-140.

Ruegg, R. and Feller, I. 2003. *A Toolkit for Evaluating Public R and D Investment*. NIST GCR 03-857. Gaithersburg, MD: National Institute of Standards and Technology.

Ruttan, V. 1982. *Agricultural Research Policy*. Minneapolis, MN: University of Minnesota Press.

Sarawetz, D. 1996. *Frontiers of Illusion* .Philadelphia, PA: Temple University Press.

Savage, J. 1999. *Funding Science in American: Congress, Universities, and the Politics of the Academic Pork Barrel*. Cambridge, UK: Cambridge University Press.

Schmoch, U., Schubert, T., Jansen, D., Heidler, R., and von Gortz, R. 2010. How to Use Indicators to Measure Scientific Performance: A Balanced Approach. *Research Evaluation,* (19): 2-18.

Schubert, T. 2009. Empirical Observations on New Public Management to Increase Efficiency in Public Research-Boon or Bane? *Research Policy,* 38:1225-1234.

Shils, E., editor. 1969. *Criteria for Scientific Development: Public Policy and National Goals*. Cambridge, MA: MIT Press.

Stokols, D., Hall, K., Taylor, B. and Moser, R. 2008. The Science of Team Science. *American Journal of Preventive Medicine,* 35: S77-S89.

U.S. House of Representatives, Committee on Science and Technology 1986. The Nobel Prize Awards in Science as a Measure of National Strength in Science, *Science Policy Study Background Report No. 3,* 99th Congress, Second Session.

Von Hippel, E. 2005. *Democratizing Innovation* . Cambridge, MA: MIT Press.

Wagner, C. and Flanagan, A. 1995. Workshop on the Metrics of Fundamental Science: A Summary, Washington, DC: Critical Technologies Institute, *Prepared for Office of Science and Technology Policy*, PM-379-OSTP.

Weingert, P. (2005) Impact of Bibliometrics upon the Science System: Inadvertent Consequences? *Scientometrics* (62): 117-131.

Wilson, J. 1989. *Bureaucracy* New York, NY:Basic Books

Zakaria, F. 2010. How to Restore the American Dream. *Time*, October 21, 2010.

APPENDIX D

THE IMPACT OF PUBLICLY FUNDED BIOMEDICAL AND HEALTH RESEARCH: A REVIEW[1]

Bhaven N. Sampat
Department of Health Policy and Management
Columbia University

I. INTRODUCTION AND BACKGROUND

New biomedical technologies trigger a number of major challenges and opportunities in health policy. Among economists, there is widespread consensus that new technologies are the major drivers of increased healthcare costs but at the same time a major source of health and welfare improvements (Murphy and Topel 2003). This has led to discussion about whether technological change in medicine is "worth it" (Cutler and McClellan 2001). The impact of new technologies on the health care system has also been the subject of much debate among health policy scholars more generally (Callahan 2009).

Public sector research agencies have an important role in the U.S. biomedical innovation system. In 2004, federal agencies funded roughly one-third of all U.S. biomedical R and D (Moses et al. 2005). The National Institutes of Health (NIH) accounted for three-quarters of this amount. Private sector drug, biotechnology, and medical device companies provide the majority of U.S. biomedical R and D funding (about 58 percent). This private sector research is, in general, focused more downstream and tends to be closer to commercial application than NIH-funded research.

---

[1] I thank Pierre Azoulay, and participants in the National Academies' 2011 Workshop on Measuring the Impacts of Federal Investments in Research, for useful comments and suggestions.

Donald Stokes (1997) observes that the public values science "not for what it is but what it is for." A perennial question in U.S. science and technology policy is what benefits taxpayers obtain from publicly funded biomedical research. Recent concerns about the clinical and economic returns to NIH funding in the post-doubling era reflect this emphasis.

In this paper, we review the evidence on the effects of publicly funded biomedical research. Reflecting Stokes's observation above, the review will focus on the health and economic effects of public research, rather than measures of scientific outcomes. Given the prominence of the NIH in funding this research, many of the published articles and research focus on this agency. The evidence examined includes quantitative analyses, and qualitative case studies, published by scholars from a range of fields. While we have made efforts to be broad, the references discussed should be viewed as representative rather than exhaustive. This review takes stock of the empirical methodologies employed and the types of data used; it also highlights common research and evaluation challenges, and emphasizes where existing evidence is more, or less, robust.

We proceed as follows. In Section II, below, we discuss a stylized model of how public research funding affects health, economic, and intermediate outcomes. As Kline and Rosenberg (1986), Gelijns and Rosenberg (1994), and others have emphasized, the research process cannot be reduced to a neat, linear model. While we recognize this fact (and highlight it in our literature review) the simple model is still useful in helping to organize our discussion of theory and data on the effects of publicly funded research. In Section III, we discuss the empirical evidence. In Section IV, we discuss common evaluation difficulties. In Section V, we conclude. The empirical approaches, data sources, and findings of many of the studies reviewed are also summarized in Tables D1–D3.

II. PUBLIC SECTOR RESEARCH AND OUTCOMES: AN OVERVIEW

Figure D-1 is a simple model illustrating how the literature has conceptualized the health and economic effects of publicly funded biomedical research (and publicly funded research more generally):

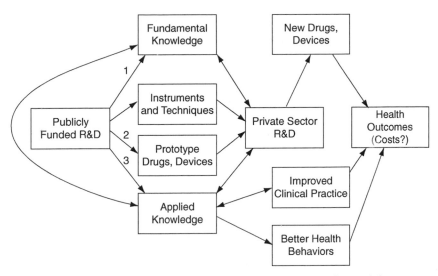

FIGURE D-1 Publicly Funded R and D and Outcomes, Logic Model
SOURCE: Sampat, 2011

The top arm of the model illustrates one important relationship: publicly funded R and D yields fundamental knowledge, which then improves the R and D efficiency of private sector firms, yielding new technologies (drugs and devices) that improve health outcomes.[2] This conceptualization has been the essential *raison-d'etre* for the public funding of science since Vannevar Bush's celebrated postwar report, *Science, The Endless Frontier*. For example, Bush asserted in 1945 that "discovery of new therapeutic agents and methods usually results from basic studies in medicine and the underlying sciences" (Bush 1945). It is also the essential mechanism in several important economic models of R and D (e.g. Nelson 1984). Importantly, this conceptualization generally views publicly funded research as "basic" research that is not oriented at particular goals, and thus yields benefits across fields. The influential "market failure" argument for public funding of basic research is that profit-maximizing, private-sector firms will tend to underinvest in this type of fundamental, curiosity driven research, since they cannot appropriate its benefits fully (Nelson 1959, Arrow 1962).

The channels through which publicly funded basic research might influence private sector innovation are diverse, including dissemination via publications, presentations and conferences, as well as through informal networks (Cohen et al. 2002). Labor markets are another

---

[2] Stokes (1997) and others have challenged this definition of "basic" research.

channel, since public agencies may also be important in training doctoral and post-doctoral students who move on to work for private sector firms (Scherer 2000).

The second arrow illustrates another relationship. New instruments and techniques that are by-products of "basic" research can also improve private sector R and D (Rosenberg 2000). Prominent examples of instruments and research tools emanating from academic research include the scanning electron microscope, the computer, and the Cohen-Boyer recombinant DNA technique.

Third, publicly-funded researchers sometimes develop prototypes for new products and processes. Some of these are indistinguishable from the informational outputs of basic research discussed above. For example, when academic researchers learned that specific prostaglandins can help reduce intraocular pressure this discovery immediately suggested a drug candidate based on those prostaglandins, though the candidate required significant additional testing and development. (This academic discovery later became the blockbuster glaucoma drug, *Xalatan*.) The public sector has also been important in developing prototypes (Gelijns and Rosenberg 1995). Roughly since the passage of the Bayh-Dole Act, in1980, publicly funded researchers have become more active in taking out patents on these inventions and prototypes for new products and processes, and licensing them to private firms (Mowery et al. 2004. Azoulay et al. 2007).

While much of the discussion of publicly funded biomedical research focuses on this more "basic" or fundamental research the public sector also funds more "applied" research and development.[3] For example, about one-third of the NIH budget is for clinical research, including patient oriented research, clinical trials, epidemiological and behavioral studies, as well as outcomes and health services research. Such research can be a useful input into the development of prototypes, and may also directly inform private sector R and D. Clinical research may also directly affect health behaviors. For example, knowledge from epidemiological research about cardiovascular health risk factors contributed to reductions in smoking and better diets (Cutler and Kadiyala 2003). New applied knowledge can also influence physicians:

---

[3] Stokes (1997) provides a thoughtful critique of conventional distinctions between "basic" and "applied" research. Since much of the literature before and since Stokes uses this terminology, we employ it in our review of this literature, even while recognizing the importance of his argument.

for example, by changing their prescribing habits (e.g. "beta-blockers after heart attacks improve outcomes") or routines (e.g. "this type of device works best in this type of patient"). Importantly, as various studies we review below will emphasize, negative results from clinical trials—showing that particular interventions do not work — can also be important for clinical practice and in shaping health behaviors.

While the discussion above assumes that new biomedical knowledge and technologies improve health outcomes, this is a topic of debate. The conventional wisdom is that while other factors (e.g. better diet, nutrition, and economic factors) were more important for health outcomes historically (McKeown 1976), improvements in American health in the post-World War II era have been driven largely by new medical knowledge and technologies (Cutler, Deaton, and Lleras-Muney 2006). The contribution of publicly funded research to these developments is an open empirical question, discussed below.

At the same time, some scholars suggest that we may have entered an era of diminishing returns, where new technologies are yielding increasingly less value (Callahan 2009; Deyo and Patrick 2004). The effect of new biomedical technologies on healthcare costs is a related concern. There is general agreement among health economists that new medical technologies are the single biggest contributor to the increase in long-run health costs, accounting for roughly half of cost growth (Newhouse 1992). Rising health costs strain the budgets of public and private insurers as well as employers, and may also contribute to generate health inequalities. The dynamic that exists between new medical technologies and health costs in the U.S. may reflect a "technological imperative," which creates strong incentives for the healthcare system to adopt new technologies once they exist (Fuchs 1995; Cutler 1995). It may also reflect positive feedbacks between demand for insurance and incentives for innovation (Weisbrod 1991).

Concern about the effects of technology on health costs has fueled empirical work on whether technological change in medicine is "worth it." Long ago, Mushkin (1979) noted (though did not share) "widespread doubt about the worth of biomedical research given the cost impacts."

A large literature in health economics suggests that new biomedical technologies are indeed, in the aggregate, worth it. Cutler (1995) and others suggest that, given the high value of improved health (current estimates suggest the value of one additional life year is $100,000 or more), even very costly medical technologies pass the cost-benefit test.[4] Nordhaus (2003) estimates that the value of improvements in health over the past half century are equal in the magnitude to measured improvements in all non-health sectors combined. Others (Callahan 2009) view these health cost increases as unaffordable, even if they deliver significant value, and therefore ultimately unsustainable.

At the same time, not all medical technologies necessarily increase costs. As Cutler (1995) and Weisbrod (1991) indicate, technologies that make a disease treatable but do not cure it - moving from non-treatment to "halfway" technology in Lewis Thomas's characterization-are likely to increase costs. The iron-lung to treat polio is an example of this. However, technologies that make possible prevention or cure ("high technology") can be cost-reducing, especially relative to halfway technologies. Thus the polio vaccine was much cheaper than the iron lung. Consistent with this, Lichtenberg (2001) shows that while new drugs are more expensive than old drugs, they reduce other health expenditures (e.g. hospitalizations). Overall, he argues, they result in net decreases in health costs (and improve health outcomes).[5]

As Weisbrod (1991) notes, "The aggregate effect of technological change on health care costs will depend on the relative degree to which halfway technologies are replacing lower, less costly technologies, or are being replaced by new, higher technologies. "[6] One way to think about the effects of public sector spending on costs would be to assess the propensity of publicly funded research to generate (or facilitate the creation of) these different types of technologies. However, since the effects of these new technologies are mediated by various facets of the health care and delivery system, it may be difficult conceptually (and empirically) to isolate and measure the effects of public sector spending on overall health costs (Cutler 1995).[7]

---

[4] Cutler (1998) observes "Common wisdom suggests that rapid cost increases are necessarily bad. This view, however, is incorrect. Cost increases are justified if things that they buy (increases in health) are worth the price paid." (2)

[5] See however, Zhang and Sourmerai (2007) for a critique of this finding.

[6] The cost-effectiveness of these technologies also depends on the populations on which they are used, as Chandra and Skinner (2011) emphasize.

[7] There is also some discussion about whether the public sector should be paying attention to the cost-side consequences of its investment decisions. Weisbrod (1991) notes: "With respect to the NIH, it would be useful to learn more about the way the size and allocation of the scientific research budget are influenced,

III. THE EFFECT OF PUBLICLY FUNDED RESEARCH: A REVIEW OF THE EVIDENCE

Health

Measuring the health returns to publicly funded medical research has been a topic of interest to policymakers for decades. In an early influential study, Comroe and Dripps (1976) consider what types of research (basic or clinical) are more important to the advance of clinical practice and health. The authors rely on interviews and expert opinion to determine the top ten clinical advances in the cardiovascular and pulmonary arena, and identified 529 key articles associated with these advances. They coded each of the key articles into six categories: (1) Basic research unrelated to clinical problems; (2) Basic research related to clinical problems (what Stokes later termed "use-oriented" basic research); (3) Research not aimed at understanding of basic biological mechanisms; (4) Reviews or syntheses; (5) Development of techniques or apparatuses for research; and (6) Development of techniques or apparatuses for clinical use. The authors find that 40 percent of the articles were in category 1, and 62 percent in categories 1 or 2. Based on this, the authors assert "a generous portion of the nation's biomedical research dollars should be used to identify and then to provide long-term support for creative scientists whose main goal is to learn how living organisms function, without regard to the immediate relation of their research to specific human diseases." Comroe and Dripps also note "that basic research, as we have defined it, pays off in terms of key discoveries almost twice as handsomely as other types of research and development combined" (1976).

A more recent set of studies examines the effects of publicly funded research on health outcomes. Operationalizing the concept of "health" is notoriously difficult. Common measures employed to account for both the morbidity and mortality effects of disease include quality adjusted life years (QALYs) and disability adjusted life years (DALYs) (Gold et al, 2002). However, it is difficult to get longitudinal information on these measures by disease. As a result, most of the analyses of the effects of public funding on health examine more blunt outcomes, including the number of deaths and mortality rates for particular diseases.

---

perhaps quite indirectly, by the health insurance system, through its impact on the eventual market for new technologies of various types" (535).

Numerous prominent academic studies (Weisbrod 1983, Mushkin 1979) aim to examine the health effects of biomedical research, and the economic value of this impact, in a cost-benefit framework. One important recent study in this tradition, Cutler and Kadiyala (2003), focuses on cardiovascular disease—the disease area where there has been the strongest improvement in health outcomes over the past sixty years. Since 1950 mortality from cardiovascular disease decreased by two-thirds, as Figure D-2 (reprinted from their paper) shows:

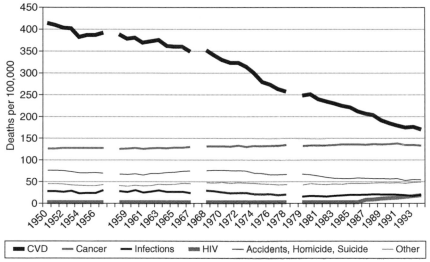

FIGURE D-2 Mortality by cause of death 1950-1994
SOURCE: Cutler and Kadiyala 2003

Cutler and Kadiyala, through a detailed review of the causes of this advance (relying on epidemiological and clinical data, medical textbooks, and other sources), estimate that roughly one third of this cardiovascular improvement is due to high-tech treatments, one third to low tech treatments, and one third to behavioral changes. Assuming one additional life year gained is valued at $100,000, the authors compute a rate of return of 4-to-1 for investments in treatments and 30-to-1 for investments in behavioral changes. These investments include costs borne by consumers and insurers, and estimates of public sector R and D for cardiovascular disease.

Based on these figures, the authors argue that the rate of return to public funding is high, though they don't directly trace public funding to changes in outcomes in their quantitative analyses. Interestingly, in their qualitative account, the major public sector research activities highlighted have an "applied" orientation, including the NIH's role in sponsoring large epidemiological trials and holding consensus conferences. This may reflect a traceability and attribution problem, which is common to the evaluation of fundamental research: It is difficult to directly link improvements in outcome indicators to public sector investments in basic research, even in a study as detailed as this one.

A paper by Heidenreich and McClellan (2003) is similarly ambitious, looking at sources of advance in the treatment of heart attacks. The authors focus on this treatment area, not only because of the large improvements, but also because it is a "best case" for attributing health outcomes to particular biomedical investments. Specifically, these authors go further than Cutler and Kadiyala by attempting to link changes in clinical practice to changes in specific R and D inputs. The authors focus here on clinical trials, not basic research. This is not because they believe that basic research is unimportant, "but because it is much easier to identify connections between these applied studies and changes in medical care and health."

Based on detailed analyses of MEDLINE-listed trials and health outcomes, the authors argue that medical treatments studied in these trials account for the bulk of improvement in AMI outcomes. The authors associate changes in clinical practice and outcomes to research results reported in trials through analysis of timing of events, and detailed clinical knowledge of how the trial results, clinical practices, and health outcomes relate.

One interesting result from this paper is that clinical practice often "leads" formal trials, challenging the "linear" model embodied in Figure D-1 (above). The authors also emphasize that an important role for trials is negative: telling clinicians what doesn't work, and stopping the diffusion of ineffective technologies. While the sample they examine represents a mix of publicly funded and privately funded trials, the authors do emphasize a particularly important role for the public sector in funding trials on drugs off patent, where private firms have fewer incentives to do so.

Philipson and Jena's (2005) study of HIV-AIDS drugs is another paper that examines the value of increases in health from new medical technologies. Though this study does not explicitly focus on the role of the public sector, it estimates that HIV-AIDS drugs introduced in the 1990s generated a social value of $1.4 trillion, based on the value of the increments to life expectancy created from these drugs (here again, using the estimate of $100,000 per life year). This study is relevant because of the important role of public sector research in the development of HIV-AIDS drugs, which is observed in several of the empirical studies discussed below.

A recent paper by Lakdawalla et al (2011) employs a similar approach to assess the benefits from cancer research. The authors find these benefits to be large, estimating the social value of improvements from improvements in life expectancy during the 1988-2000 period to be nearly $2 trillion. The authors note that this compares to investments of about $80 billion dollars in total funding for the National Cancer Institute between 1971 and 2000. As with the HIV studies discussed above, the authors do not calculate a rate-of-return on publicly funded research explicitly, but do argue that the social benefits from cancer research in general far exceed research investments and treatment costs.

A large share of the benefits in the cancer arena, according to this work, results from better treatments. Lichtenberg (2004) also suggests that new drug development has been extremely important in progress against cancer.[8] Public sector research may have been important to the development of these drugs: various studies (Stevens et al. 2011, Chabner and Shoemaker 1989) suggest an important role for the public sector in cancer drug development.[9]

Each of the studies discussed so far focuses on particular disease areas. In a more "macro" approach Manton (2009) and colleagues relate mortality rates in four disease areas to lagged NIH funding by the relevant Institute, over the period 1950-2004. They find that for two of the four diseases (heart disease, stroke) there is a strong negative

---

[8] Cutler (2008) also emphasizes progress in the "war on cancer" – though highlights the role of screening and personal behavior changes, and notes the high costs of treatment. Sporn (2006) and Balilar and Gonik (1997) offer less sanguine assessments, emphasizing that progress against cancer has been highly uneven. Long-standing debates in assessments of the War on Cancer include the disagreements on the relative importance of treatment versus prevention, and of basic versus applied research. The literature also suggests it is difficult to evaluate the extent of progress in cancer, for two main reasons. First, advances in screening increase incidence. The second is competing risks: for example, the reduction in mortality from cardiovascular disease, discussed above, increased cancer cases. See Cutler (2008) for a review.

[9] A National Cancer Institute (NCI) "Fact Sheet" asserts that "approximately one half of the chemotherapeutic drugs currently used by oncologists for cancer treatment were discovered and/or developed at NCI."
http://www.cancer.gov/cancertopics/factsheet/NCI/drugdiscovery

correlation, but find weaker evidence for cancer and diabetes. Several issues arise here that will re-emerge in other quantitative analyses discussed below. First, linking funds to disease areas is difficult. As with other studies we will consider below, the authors here rely on the disease foci of Institutes within the NIH. More importantly, the counterfactual is hard to prove: It is difficult to make the case that the relationships estimated are causal, since Institute-specific funding is not exogenous. In particular, diseases where there is highest expectation of progress (even absent funding) may be more likely to get funds. Finally, competing risks also complicate interpretation of health outcomes. For example, part of the reason cancer mortality has increased rather than decreased over the period studied is that people no longer die of heart attacks, due to advances in the cardiovascular arena.

Private Sector R and D

Another set of studies relates publicly funded research to private sector R and D and productivity. These include econometric analyses relating public sector and private sector funding, surveys of firm R and D managers, and studies examining the geographic dimension of spillovers from public sector researchers.

Several papers relate NIH funding by disease area to later private sector funding. One motivation in these studies is to assess if public and private sector R and D are substitutes or complements, an issue of perennial interest in science and technology policy (David, Hall, and Toole 2000). The econometric analyses generally find a positive association between public sector and private sector funding. Toole (2007) uses data from the NIH'S Computerized Retrieval of Information on Scientific Projects (CRISP) database, covering NIH basic and clinical research funding across seven therapeutic classes (between 1972 and 1996), and data from the Pharmaceutical Manufacturers of America (PhRMA) on private sector R and D in these same areas (between 1980 and 1999) to examine the relationships between the two. This study finds a 1 percent increase in basic research funding associated with a 1.7 percent increase in private sector funding, though the elasticity for clinical research is much smaller (.4 percent). In a similar analysis, Ward and Dranove (1995), using PhRMA data on R and D spending and NIH data on funding by Institute (similar to that used in the Manton et al 2009 study discussed above) find that a 1 percent increase in NIH research support in a disease area is associated with a .76 percent increase in private sector R and D within that same disease area over the next seven years.

Surveys of firm R and D managers have also been used to gauge how public sector research affects private sector R and D. Cohen, Nelson, and Walsh (2002) report on the 1994 Carnegie Mellon Survey of Industrial R and D managers, which examined (among other issues) the roles of the public sector in industrial R and D, and channels through which public research affects industrial R and D. This survey is particularly interesting since it has data on both the drug and device sectors, and allows for comparison of these sectors to others. The authors find that the pharmaceutical industry is an outlier in its reliance on public sector R and D. In the pharmaceutical industry, according to respondents, public research was the most important source of new project ideas *and* contributor to project completion. By contrast, in the medical instruments industry R and D projects less frequently rely on public research than other industries. There are also some differences in the fields of science relied upon across these different industries. Thus the top three fields of science important to R and D projects in the pharmaceutical industry are medicine, biology, and chemistry. In medical instruments sector, the top three fields are medicine, materials science, and biology. Although much of the literature on the effects of public sector funding tend to focus on the NIH, the bulk of funding for materials science R and D comes from other agencies (including the National Science Foundation, Department of Energy, and the Department of Defense).

Another set of studies, examining how interactions between public and private sector scientists affects the productivity of private sector R and D, generally finds a strong relationship between the two. Cockburn and Henderson (1996) examine how private sector co-authorship with public sector scientists affects firm level R and D and productivity. The authors bring together several novel datasets, including MEDLINE data on firm publication activity and USPTO data on firm patenting activity. Using panel regression models (with firm fixed effects to control for time-invariant firm characteristics), they find a positive and statistically significant association between their productivity measure (based on important patents per R and D dollar) and collaboration with public sector scientists.

Research by Zucker, Darby, and Brewer (1998) examines the importance of academic science in the creation of new biotechnology firms in the 1980s. In this work, the authors relate new biotechnology firm formations by area to the number of academic "star scientists" (as measured by publications and other measures of scientific productivity) working in that area. The authors find that the presence of academic stars and their collaborators— intellectual capital"—within a geographic area has a statistically significant and positive relationship with the number of new biotechnology enterprises later formed in that area. This research suggests that public sector science has an important, though geographically mediated, effect on private sector research.

The question of whether spillovers from public research to firms are geographically mediated has also been examined through studies using patent citation data (Jaffe et al. 1993). When patents are granted they include citations to prior art: earlier publications and patents that were deemed (by either the applicant or the patent examiner) as relevant to an invention. Economists and others have interpreted patent citations as evidence of knowledge flows or spillovers: thus if a firm patent cites a public sector publication or patent, this is considered evidence that the firm benefited from public funding. While there is some skepticism about this measure, given the prominence of patent examiners in generating citations (Alcacer et al. 2009; Cohen and Roach 2010), it remains commonly employed. Moreover, as it turns out, examiner-added citation are less common within the biomedical arena (Sampat 2010) and for citations to scientific publications (Lemley and Sampat 2011) suggesting that citations in biomedical patents to scientific publications may be less subject to the concerns cited above.

Azoulay, Graff Zivin, Sampat (2011) collected data on 10,450 elite life science researchers (most of them publicly funded), historical information on productivity, employment locations of each scientist, MEDLINE data on their publications, ISI data on citations to their publications, and USPTO data on their patents and citations to their patents and publications. The authors assess the effects of geography on spillovers by examining how citation patterns change after the scientists move. Overall, they find some evidence that geography matters for spillovers, though weaker than in previous analyses. They also find the results on geography are sensitive to whether spillovers are measured through paper-to-paper citations, patent-to-patent citations, or patent-to-paper citations.

Private Sector Innovation

Numerous studies also consider the public sector role in the development of marketed innovations. Survey work by Mansfield (1998) examines the importance of academic research for industrial innovation for firms across a range fields. In this work, as in the Carnegie Mellon Survey discussed above, the biomedical industries are outliers. The share of products developed over the late 1980s and early 1990s that could *not* have been developed (without substantial delay) absent recent academic research is nearly twice as high in drugs and medical products than in other industries.

Various recent studies examine the roles of the public sector in drug development using patent and "bibliometric" data. In addition to providing an indicator of returns to public R and D, this work may also be relevant to current policy proposals that aim to exploit public sector ownership of drugs to help reduce downstream drug prices and expand access (Sampat and Lichtenberg 2011).

Sampat (2007) uses data on all drugs approved by the Food and Drug Administration (FDA) between 1988 and 2005 (and listed on the FDA's *Orange Book*), and USPTO data on patents associated with these drugs, to examine the share of drugs on which academic institutions (including public sector laboratories) own patents. Overall, a small number of new molecular entities (NMEs), about 10 percent, have academic patents. However, this share is larger for new molecular entities that received priority review (arguably the most innovative new drugs), where about 1-in-5 drugs have academic ownership. He also finds that public sector ownership of drugs is more pronounced for HIV-AIDS drugs than for other drug classes.

Stevens et al. (2011) expand on this research to include vaccines and biologicals (not always listed on the Orange Book), and construct measures based not only on publicly available patent data but also propriety data on drug licenses. They find 153 FDA-approved drugs were discovered by the public sector over the past 40 years (102 NMEs, 36 biologics, and 15 vaccines.) The authors show that about 13 percent of NMEs (and 21 percent of priority NMEs) were licensed from public sector institutions, consistent with the numbers reported in Sampat (2007). Strikingly, the authors also show that virtually all the important vaccines introduced over the past quarter century came from the public sector. The authors also show broad correlations between NIH Institute budgets and the therapeutic classes where there are numerous public-sector based drugs, similar in spirit to econometric analyses we will review below.

Kneller (2010) takes a different approach, relying not on patent assignment records but instead on information related to the inventors' places of employment, and applies his analysis to 252 drugs approved by the FDA between 1998 and 2007. Using these measures, Kneller finds a larger public sector influence than the previous studies. Overall, about a quarter of drugs are from university inventors, and a third of priority review drugs are from academic inventors.

The Sampat, Stevens et al, and Kneller studies rely on direct academic involvement in developing the molecules (resulting in academic ownership of the key patents or academic inventors listed on those patents). However, as discussed in Section II, in addition to the development prototypes, the public sector can facilitate or enhance industrial innovation in other ways as well. Thus Keyhani et al (2005), using data from the Federal Register, government clinical trials databases, and documents from the FDA, finds the government was active in supporting clinical trials for nearly 7 percent of a sample of drugs approved between 1992 and 2002. Here again, the government role was more pronounced for HIV-AIDS drugs than for others.

Sampat and Lichtenberg (2011) distinguish between the direct effects of public sector research on drug development, where academic institutions are involved in discovering the molecule, and the indirect effects, where other knowledge spillovers from academic work increase private sector productivity. The authors measure the direct effect of public sector funding using information on "government interest" statements in Orange Book listed patents. And they use citations in Orange Book listed patents to academic patents or academic publications as a measure of this indirect effect. Consistent with the various studies cited above, this study suggests the direct effect is small overall: about 9 percent of drugs, and about 17 percent of priority review drugs, have public sector owned patents. However the indirect effect is much larger: about 48 percent of drugs have patents that cite public sector patents and publications. Among priority drugs, this indirect influence rises to nearly two-thirds. This finding is broadly consistent with the qualitative results from Cockburn and Henderson's (1996) study of fifteen drugs, which shows the public sector made key enabling discovery for the majority (11 of the 15), but was involved in synthesis of the compound for only 2 of the 15.

The studies discussed above are accounting exercises. Others also have attempted to relate variation in funding by disease area to drug development patterns, econometrically. Dorsey et al. (2009) relate NIH funding by therapeutic area to later drug approvals across nine disease areas between 1995 and 2000. The authors allocate funding to specific diseases based on funding Institute using information in Congressional budget requests for those institutes. They find that despite a sharp rise in NIH funding over this time period, drug approvals remained flat overall. And their cross-therapeutic area analyses show little correlation between NIH funding and subsequent drug approvals.

Blume-Kohut (2009) also explores these issues, using panel regression models. She constructs data on NIH funding by disease area between 1975 and 2004 from the agency's CRISP and RePORTER databases, based on parsing of abstracts and keywords of grants for disease keywords. She also examines information on drugs in development by class using data from a private data vendor, PharmaProjects. Her results show little evidence of responsiveness between the number of drugs in Phase III trials (late stage) and NIH funding, but evidence of a positive relationship for the number of drugs in earlier stage Phase I trials. The author notes these results may suggest that factors other than NIH funding (or the state of knowledge) may be important for Phase III trials, including commercial considerations such as the size of the market. In a similar approach, using a different outcome measure, Ward and Dranove (1995) relate MEDLINE publications tagged as "drug" articles to NIH R and D funding by disease area, here again categorized based on funding institute. They find a strong relationship between the two.

Most of the studies we have discussed thus far, examining public sector research and product development, focus on drugs and involve quantitative analysis. By contrast, Morlacchi and Nelson (2011) examine the sources of innovation in the development of the left ventricular assist device (LVAD), a medical device used for patients with end-stage heart failure. While the device originally was developed as a "bridge" solution until a heart became available for transplant, it is increasingly used as destination therapy, as a substitute for a heart transplant. Morlacchi and Nelson draw on interviews, primary and secondary articles, and patents to develop a longitudinal history of the development of the LVAD. They consider, among other questions, the importance of public sector funding in this development. Echoing some of the themes in Heidenreich and McClellan's study of heart attack treatment, they find that in this field application led scientific understanding. The development of the device occurred even as basic understanding of heart failure remained weak, once again challenging the linear model of innovation portrayed in Figure D-1. They also find that the applied and diffusion oriented activities of public sector funders were important in the development of this device, including the NIH's sponsorship of conferences and centers to spread best-practice, funding of trials and development of important component technologies, and contracts to spur firm formation.

Health Costs

Despite longstanding concerns about the effects of new biomedical technologies on healthcare costs, and speculation that public sector research may be implicated in spurring this cost spiral, there has been surprisingly little empirical research on this topic. For example, there is a paucity of academic work relating funding patterns by disease area to subsequent cost growth, analogous to the work relating funding to private sector R and D, drug development, and health outcomes discussed above.

In 1993, the NIH prepared studies on the cost savings from a non-random sample of 34 health technologies resulting from NIH support, demonstrating substantial cost savings (NIH 1993). This study examined NIH funding for new technologies, as well as cost savings that accrued to patients, based on conservative assumptions on reductions in disease attributable to those same technologies. An NIH summary (NIH 2005) of this work notes that,"[t]aken together, the 34 technologies were estimated to reduce health care costs by about $8.3 billion to $12.0 billion annually." As with several studies discussed earlier, difficulty in tracing the effects of "basic" research to particular technologies may complicate such calculations. Moreover, as the agency's summary emphasizes "because the 34 new health care technologies studies were not chosen to be representative of all health advances resulting from NIH support, the results of these case studies cannot be generalized."

While there has been little work, beyond this NIH study, on the effects of public sector funding on the direct costs of disease (i.e. health expenditures), the various studies discussed above that address the value of new biomedical technologies, can be interpreted as evidence that public sector funding reduces the total cost of disease, to the extent that the estimated improvements in health are viewed as reductions in the social costs associated with disease.

III. MEASUREMENT AND EVALUATION ISSUES

The diverse set of studies reviewed here illustrates a number of common measurement and evaluation issues that complicate efforts to estimate the health and economic effects of publicly funded biomedical research. Here, we will highlight several that stand out.

Several of the studies reviewed relate public sector funding by disease area to outputs. All of these focus on the NIH, since for other agencies publicly available data on funding by disease area is not readily available. Even for the studies focused on the NIH, however, there are measurement issues. While many studies construct funding stocks based on which Institutes fund the research, Institutes fund numerous diseases, introducing considerable noise into these measures.

The NIH's CRISP database includes disease keywords, which can also be used to construct disease specific funding, but these are not collected in a standard way across the NIH (Sampat 2011). In 2008, the NIH launched the "Research, Condition, and Disease Categorization" (RCDC) database, which uses standard methodologies to classify funds by area. Whereas, previously, each NIH Institute had linked its grants to diseases in an ad hoc and non-standard way, the RCDC employs standard category definitions to classify grants, developed with input from disease groups, the scientific community, and outside consulting groups. Before the RCDC, the NIH had provided disease-specific funding figures tentatively and with many caveats. Today, with the existence of the RCDC database, the agency has exhibited a more firm commitment to its own data sources and tracking. The NIH website thus affirms: "RCDC provides consistent and transparent information to the public about NIH-funded research. For the first time, a complete list of all NIH-funded projects related to each category is available." This database may prove a boon for future researchers. However, its time frame and scope (covering only diseases and conditions "of historical interest to Congress") may limit the types of analyses that can be conducted using these data.

A more fundamental issue is difficulty in categorizing "basic" research in these studies. Thus in the CRISP funding database, 49% of grants awarded in 1996 (accounting for 46% of NIH allocations) listed no disease terms, and only about 45% of grants map to a disease category in the RCDC (Sampat 2011). It is difficult to incorporate these grants into disease level associations of funding and outputs. Basic research is also difficult to trace to outcomes even in a case study context, given lags and diffuse channels of impact. Thus it is not surprising that several of the evaluation studies discussed above (including the study of heart attack treatment, and the studies of NIH research and costs) focus on the effects of applied research.

The bibliometric approaches discussed above, linking grants to publications to citations to patents to drugs may overcome these traceability challenges, relying on paper trails between research and outcomes, and avoiding the need to associate public sector funding with particular diseases. However, the validity of these analyses rest on a number of assumptions, e.g. the extent to which patent-paper citations reflect real knowledge flows from public sector research.

Thus, measurement of inputs and intermediate steps is difficult. Measuring outcomes is conceptually easier, at least relative to evaluation of research outputs in non-biomedical contexts. Though the right output measures (e.g. morbidity or mortality, direct or indirect costs) or desiderata (should the NIH be mainly focused on advancing health? science? competitiveness? something else?) are the subject of debate, there is a wealth of data available to examine changes in health-related outcomes. Similarly, the research community has exploited numerous useful measures of relevant economic outcomes (e.g. patents, drug development, publications), again more readily available in the biomedical context than other arenas.

Causal evaluation of the effects of publicly funded research on these outcomes is difficult however, in this context and in S and T policy more generally. Simply put, funding choices are not random, making it difficult to attribute observed changes in outcomes to specific policies. As just one example, if public sector funding targets disease areas with high scientific opportunity, it is difficult to untangle whether subsequent improvements in health (or changes in private sector R and D, or drug development) reflect the effects of the funding or of the scientific opportunity. Several of the studies discussed attempt to address this problem econometrically, including through panel regression models with disease fixed effects, to absorb the effects of disease-specific characteristics that do not change over time. Going forward, quasi-experimental techniques may also prove useful. For example, it may be possible to exploit random shocks to funding in particular areas that are unrelated to scientific opportunity and disease burden could (e.g. those introduced through political influence on the allocation process, or changes in agencies' funding rules) to assess the effects of public research.

There is also a need for more qualitative work. A number of the case studies surveyed above relied on detailed knowledge of the institutions at play, in depth clinical knowledge, and information on the timing of relevant events, to make credible arguments that the relationships they observed were causal. These too represent promising research approaches going forward.

IV. CONCLUSIONS

The measurement evaluation challenges highlighted above are endemic to science and technology policy in general (Jaffe 1998). A main output of science and technology policy is knowledge, which is difficult to measure and link to downstream outcomes. This exacerbates traditional difficulties with attributing causal effects to policy interventions, common to evaluation in most public policy domains.

Notwithstanding these challenges, at least on several issues various studies point in the same direction. First, there is consistent evidence across on the importance of public sector biomedical R and D for the efficiency of private sector R and D. The evidence is compelling since it is based on a range of studies using different techniques and samples, including surveys, case studies, and econometric analyses.

Second, the accounting studies on sources of innovation in drugs suggest that the public sector was directly involved in the development of a small share of drugs overall, but that the public sector role is more pronounced for more "important" drugs, and that the indirect effect of public sector research on drug development is larger than the direct effect. On the other hand, the studies that relate patterns of funding by disease area to drug development show less consistent results.

Third, a number of the studies suggest the importance of the applied and clinical public research activities on product development, patient behaviors, and health outcomes. This is striking, since much of the discussion about publicly funded biomedical research focuses on (and most of the funding is for) "basic" research. Whether the importance of applied activities reflects that their effects are easier to measure and trace, or that they are really very important, is an open empirical question.[10]

Overall, there is strong evidence that new biomedical technologies have created significant value, as measured through the economic value of health improvements. Some scholars believe that even if public sector

---

[10] However, recall that the Toole (2007) study shows that basic research funding by the public sector has a stronger effect on private R and D than clinical research funding.

research was responsible for only a small share of this gain, it delivers high returns on investment (Murphy and Topel 2003).[11]

More work is needed directly examining the role of the public sector per se, and especially public sector basic research, in affecting these health outcomes. Similarly, very little is known about the effects of public sector research on health expenditures. Detailed longitudinal case studies of trends in public and private sector research activity, technology utilization, health outcomes, and health expenditures across a number of disease areas would be useful for promoting understanding on each of these issues. To the extent possible, it would be useful for these studies to employ common methods and measures, and to examine both disease areas where there has been considerable advance, and those where there has been less progress.

Finally, the bulk of the academic work in this area focuses on the NIH and pharmaceuticals. Much more research is needed on the effects of other funding agencies, and on the effects of public funding on the device sector.

REFERENCES

Alcacer, J., Gittleman, M., and Sampat, B.N. 2009. Applicant and Examiner Citations in U.S. Patents: An Overview and Analysis. *Research Policy*. 38(2): 415-427.

Arrow, K. 1962. Economic Welfare and the Allocation of Resources for Invention in Richard Nelson ed. *The Rate and Direction of Inventive Activity*, Princeton, NJ: Princeton University Press.

Azoulay, P., Michigan, R., and Sampat, B.N. 2007. The Anatomy of Medical School Patenting. *The New England Journal of Medicine* 20(357): 2049-2056.

Azoulay, P., Zivin, J.G., and Sampat, B.N. 2011. The Diffusion of Scientific Knowledge across Time and Space: Evidence from Professional Transitions for the Superstars of Medicine. *NBER Working Paper 16683*.

---

[11] Heidenreich and McClellan (2003) summarize this point of view in the introduction to their study (discussed above), noting that while previous analyses "have generally not provided direct evidence of the impact on health of specific research studies, or on the likely value of additional research funding" these previous studies tend to conclude "recent gains in health are extraordinarily valuable in comparison with the relatively modest past funding."

Ballar, J. and Gornick, H. 1997. Cancer Undefeated. *New England Journal of Medicine. 336*: 1569-1574.

Blume-Kohout, M. E. 2009. Drug Development and Public Research Funding: Evidence of Lagged Effects. 1-35. Waterloo, ON, Canada: University of Waterloo.

Bush, V. 1945. *Science,The Endless Frontier.* Washington, DC: United States Government Printing Office.

Callahan, D. 2009. Taming the Beloved Beast: How Medical Technology Costs Are Destroying Our Health Care System. Princeton: Princeton University Press.

Chabner, B. A. and Shoemaker, D. 1988. Drug Development for Cancer: Implications for Chemical Modifiers. *International Journal of Radiation Oncology Biology Physics.* 16, 907-909.

Chandra, A., and Skinner, J. 2011. Technology Growth and Expenditure Growth in Health Care. *NBER Working Paper 16953.*

Cockburn I, H. R. 1996. Public-Private Interaction in Pharmaceutical Research. *Proceedings National Academy of Science USA* 93(23): 12725-12730.

Cohen W.M., Nelson, R., and Walsh J. 2002. Links and Impacts: The Influence of Public Research on Industrial R and D. *Management Science,* 48(1): 1-23.

Cohen, W. and Roach, M. 2010. Patent Citations As Indicators of Knowledge Flows From Public Research. *Working Paper.*

Comroe, J. and Dripps, R.D. 1976. Scientific Basis for the Support of Biomedical Science. *Science,* 192(4235): 105-111.

Cutler, D. 1995. Technology, Health Costs, and the NIH. Paper prepared for the National Institutes of Health Economic Roundtable on Biomedical Research

Cutler, D. M. 2008. Are We Finally Winning the War on Cancer? *Journal of Economic Perspectives,* 22(4): 3–26.

Cutler, D and McClellan, M. 2001. Is Technological Change in Medicine Worth It? *Heath Affairs,* 20(5): 11-29.

Cutler, D., and Kadiyala, S. 2003. The Returns to Biomedical Research: Treatment and Behavioral Effects, in *Measuring the Gains from Medical Research: An Economic Approach*, eds Kevin Murphy and Robert Topel, Chicago: University of Chicago Press 110-162.

Cutler, D., Deaton, A., and Lleras-Muney, A. 2006. The Determinants of Mortality. *The Journal of Economic Perspectives,* 20(3).

David, P. A., Hall, B.A, and Toole, A.A. 2000. Is Public R and D a Complement or Substitute for Private R and D? A Review of the Econometric Evidence. *Research Policy,* 29:497–529

Dorsey, E.R, Thompson, J.P., Carrasco, M., de Roulet, J., Vitticore, P., Nicholson, S., Johnston, S.C., Holloway, R.G., Moses, H III. 2009. Financing of U.S. Biomedical Research and New Drug Approvals across Therapeutic Areas. *PloS One 4,* 9. e7015.

Dorsey, E.R., Vitticore, P., De Roulet, J., Thompson, J.P., Carrasco, M., Johnston, S.C., Holloway, R.G., Moses H III. 2006. Financial Anatomy of Neuroscience Research. *Annals of Neurology,* 60, (6): 652-659.

Fuchs, V. 1986. *The Health Economy.* Cambridge, MA: Harvard University Press.

Gelijns, A., and Rosenberg, N. 1994. The Dynamics of Technological Change in Medicine. *Health Affairs,* 13*(*3): 28-46

Gelijns, A. and Rosenberg, N. 1995. The Changing Nature of Medical Technology Development in Rosenberg, N., Gelijsn, A., and Dawkins,H. (eds.). *Sources of Medical Technology: Universities and Industry,* Washington National Academies Press

Gold, M., Stevenson, D., and Fryback, D. 2002. HALYS and QALYS and DALYS, OH MY: Similarities and Differences in Summary Measures of Population Health. *Annual Review of Public Health* 23: 115-134.

Heidenreich, P., and McClellan, M. 2003. Biomedical Research and Then Some: The Causes of Technological Change in Heart Attack Treatment. In *Measuring the Gains from Medical Research: An Economic Approach,* edited by Kevin Murphy and Robert Topel, 163-205. Chicago: University of Chicago Press.

Jaffe, A., Trajtenberg, M., and Henderson, R. 1993. Geographic Localization of Spillovers as Evidenced By Patent Citations. *The Quarterly Journal of Economics,* 108*(*3): 577-598.

Keyhani, S., Diener-West, M., and Powe, N. 2005. Do Drug Prices Reflect Development Time and Government Investment? *Medical Care,* 43(8): 753-762.

Kline, S., and Rosenberg, N. 1986. An Overview of Innovation in Ralph Landau and Nathan Rosenberg, (eds.). *The Positive Sum Strategy: Harnessing Technology for Economic Growth.*

Kneller, R. 2010. The Importance of New Companies for Drug Discovery: Origins of a Decade of New Drugs. *Nature Reviews Drug Discovery* 9 (11): 867-882.

Lemley, M., and Sampat, B.N. 2011. Examiner Characteristics and Patent Office Outcomes. Forthcoming, *Review of Economics and Statistics.*

Lichtenberg, F. 2001. Are the Benefits of New Drugs Worth Their Cost? *Health Affairs*(20): 241-251

Mansfield, E. 1998. Academic Research and Industrial Innovation: An Update of Empirical Findings. *Research Policy* (26) 7-8:773-776.

Manton, K; Gu, X; Lowrimore, G; Ullian, A; Tolley, H. 2009. NIH Funding Trajectories and Their Correlations with Us Health Dynamics from 1950 to 2004.. *Proceedings National Academy of Science USA,* 106(27): 10981-10986.

McKeown T. 1976. *The Role of Medicine: Dream, Mirage, or Nemesis?* London: Nuffield Provincial Hospitals Trust.

Morlacchi, P,, and Nelson. R.R. 2011. How Medical Practice Evolves: The Case of the Left Ventricular Assist Device. *Research Policy.* 40(4): 511-525

Moses III, H., Dorsey, ER, Matheson, D.H.M., and Thier, S.O. 2005. Financial Anatomy of Biomedical Research. *Journal of the American Medical Association ,* 294(11): 1333-1342.

Mowery, D., Nelson, R.R., Sampat, B.N., and Ziedonis, A.A. 2004. *Ivory Tower and Industrial Innovation: University–Industry Technology Transfer Before and After Bayh-Dole.* Stanford, CA: Stanford University Press.

Murphy, K. M. 2003. *Measuring the Gains from Medical Research: An Economic Approach.* 1st ed: University of Chicago Press.

Murphy, K., Topel, R.,(eds.). 2003. *Measuring the Gains from Medical Research: An Economic Approach*, Chicago, IL: University of Chicago Press

Mushkin, S. 1979. *Biomedical Research: Costs and Benefits.* Cambridge, MA: Ballinger Publishing.

National Institutes of Health. 1993. Cost savings resulting from NIH research support : a periodic evaluation of the cost-benefits of biomedical research.

Nelson, R.R. 1959. The Simple Economics of Basic Scientific Research , *Journal of Political Economy*, (67): 297-306.

Nelson, R.R. 1982. The Role of Knowledge in R and D Efficiency. *The Quarterly Journal of Economics*, 97(3): 453-470.

Nordhaus. 2003. The Health of Nations: The Contribution of Improved Health to Living Standards in Kevin Murphy and Robert Topel, eds., *Measuring the Gains from Medical Research: An Economic Approach.*

Philipson, T., and Jena , A.B. 2006. Who Benefits from New Medical Technologies? Estimates of Consumer and Producer Surpluses for HIV/AIDS Drugs. *Forum for Health Economics and Policy*, *9* (2) (Biomedical Research and the Economy), Article 3.

Rosenberg., N. 2000. Schumpeter and the Endogeneity of Technology: Some American Perspectives. London : Psychology Press

Sampat, B.N. 2009. Academic Patents and Access to Medicines in Developing Countries. *American Journal of Public Health,* 99 (1): 9-17.

Sampat, B.N . 2010. When Do Patent *Applicants Search for Prior Art?* Journal of Law and Economics. 53:399-416.

Sampat, B.N. 2011. The Allocation of NIH Funds Across Diseases and the Political Economy of Mission-Oriented Biomedical Research. *Working paper.*

Sampat, B.N., and Lichtenberg, F. 2011. What Are The Respective Roles Of The Public And Private Sectors In Pharmaceutical Innovation? *Health Affairs*, 30(2): 332-339.

Scherer, F.M. 2000. The pharmaceutical industry, Chapter 25 in Anthony Culyer and Joseph Newhouse eds., *Handbook of Health Economics*. Amsterdam: North Holland

Sporn, M. B. 1997. The War on Cancer: A Review. *Journal of the New York Academy of Sciences*, 833(1).

Stevens,A. J., Jensen, J. J., Wyller, K., Kilgore, P. C., Chatterjee ,S., Rohrbaugh, M. L. 2011. The Role of Public-Sector Research in the Discovery of Drugs and Vaccines. *The New England Journal of Medicine,* 364(6): 535-541.

Stokes, D. 1997. Pasteur's Quadrant: Basic Science and Technological Innovation. Washingon Brookings Institution Press.

Toole, A.A. 2007. Does Public Scientific Research Complement Private Investment in Research and Development in the Pharmaceutical Industry? *The Journal of Law and Economics,* 50(1): 81-104.

Ward, M.,R., Dranove, D. 1995. The Vertical Chain of Research and Development in the Pharmaceutical Industry. *Economic Inquiry, 33*: 70-87.

Weisbrod, B. A. 1983.*Economics and Medical Research.* AEI Press.

Weisbrod, B. A. 1991. The health care quadrilemma: an essay on technological, change, insurance, quality, and cost containment. *Journal of Economic Literature* , 523-552

Zhang , Y., and Soumerai, S. 2007. Do Newer Prescription Drugs Pay for Themselves? A Reassessment of the Evidence. *Health Affairs, 26*(3), 880-886.

Zucker, L., Brewer, M., and Darby, M. 1998. Intellectual Human Capital and The Birth of U.S. Biotechnology Enterprises. *American Economic Review*, 88 (1): 290-306

.

Table D-1 Public Funding and Health Outcomes: Summary of Selected Studies

Authors	Question	Empirical Approach	Measures/Data	Results
Cutler and Kadiyala (2007)	What is the role of biomedical research in reduction in CVD mortality? What is rate of return on biomedical research funding?	Detailed case study of the roles of high tech invasive treatments, medications, behavioral changes in overall improvement Residual based approach to decompose roles of each in improvement Analyses of the roles of medical research in advancements above Estimate costs of total research Relate benefits to costs to calculate rates of return; rely on historical record for causality claims; robustness checks using alternative assumption	Economic value of clinical benefits of medical treatments, changes in behavior Data on NIH funding for cardiovascular disease 1953-1997	Returns to basic research 30-1 Much of the benefit is through effects on behavioral change (smoking etc.) which they attribute to NIH via historical record

Authors	Question	Empirical Approach	Measures/Data	Results
Weisbrod (1983)	What was rate of return on public investments in polio research?	Detailed case study. Counterfactual: what would clinical and economic costs be in absence of vaccine?	Economic value of clinical outcomes. Relate to data on public expenditures on "polio"	Rate of return 11-12%
Heidenreich and McClellan (2007)	How important has biomedical research been in care of heart attacks?	Focus on applied research "not because we view basic research as unimportant, but because it is much easier to identify connections between these applied studies in medical care and health". Decompose sources of improved outcomes for heart attack treatment over 1975-1995. Use information on timing of key trials to infer causality. Qualitative analyses relating trials to outcomes	Medline data on relevant trial, timing of major RCTs. Trends in use of interventions. 30 day mortality post-AMI. Funding sources for the trials	Mini-case studies show RCTs have some effect on clinical practice (thrombolytic drugs), but small. Most other trials had a limited effect. Negative trials had lagged but real effects. Clinical practice leads doesn't lag. Formal applied studies alone don't explain much of the decline; a lot of learning is informal

Authors	Question	Empirical Approach	Measures/Data	Results
Manton et al (2009)	How do U.S. health dynamics relate to NIH funding patterns from 1950 to 2004?	Correlate 10 year lagged NIH funding to outcomes for four major chronic diseases: CVD, stroke, cancer, diabetes	NIH funding overall (lagged 10 years) NIH funding for four relevant institutes (NHLBI, NINDS, NCI, NIDDK) Outcome measures: cause specific mortality (deaths/100,000); age adjusted death rates	Temporal correlation between funding from relevant institute and deaths for 3 of the 4 diseases Lagged NIH funding negatively correlated with age adjusted death rates for 2 of 4 diseases (heart disease, stroke) Using counterfactuals based on historical trends, project significant deaths averted due to NIH funding (mostly CVD)

Authors	Question	Empirical Approach	Measures/Data	Results
Comroe and Dripps (1976)	What types of research (clinical vs. basic) are important in the advance of clinical practice, health?	Interviews, expert opinions used to determine of top 10 clinical advances in cardiovascular and pulmonary arena Content analyses of key articles	Top 10 clinical advances "Key articles" associated with these advances Coding of whether the key articles are clinical or non-clinical	41 percent of all work judged to be essential or crucial for later clinical advances was not clinically oriented at the time of research

TABLE D-2 Public Funding and New Drugs, Devices: Summary of Selected Studies

Authors	Question	Empirical Approach	Measures	Results
Cockburn and Henderson (1996)	How does public sector research affect pharmaceutical innovation?	Case studies of 15 clinically important drugs	Qualitative determinations of roles of public sector in drug development	Of 15 drugs, public sector research made key enabling discovery for 11 Public sector involved in synthesis of major compound in 2 cases
Ward and Dranove (1995)	How do MEDLINE "drug" articles respond to NIH funding?	Panel regressions articles in a disease area to NIH R and D by relevant institute	NIH data on R and D by institute MEDLINE data on publications by disease area	Strong relationship between NIH funding and later MEDLINE articles Indirect effect (from research outside disease area) stronger than direct effect
Sampat and Lichtenberg (2011)	What are the roles of the public and private sectors in drug development?	Examine share of new molecular entities where public sector developed patent (direct effect) and where private sector patents cite public sector patents/publications (indirect effect)	FDA approved NMEs 1988-2005 Orange Book patents on these drugs Government interest statements/assignment in patents Backward citations in patents to public sector patents, MEDLINE	Direct effect: public sector owns key patent for 9% of drugs Indirect effect: Public sector patents or publications cited by 48% of drugs Both direct and indirect effects more pronounced for most clinically important drugs (17%, 65%)

Authors	Question	Empirical Approach	Measures	Results
Sampat (2007)	On how many drugs do academic institutions own patents?	Examine share of drug approvals where academic and public sector institutions own key patents	FDA approved NDAs 1988-2005 Orange Book patents on these drugs USPTO data on patent ownership Azoulay-Sampat concordance of academic assignees articles acknowledging public sector funding	72 of 1546 NDAs have an academic patent 10.3 percent of NMEs 5.9 percent of non-NMEs 19.2 percent of priority NMEs have an academic patent
Keyhani et al (2005)	Do drug prices reflect development time and government investment?	Regression analyses relating drug prices to measures of government support	180 drugs listed in the Federal Register between 1992 and 2002 Federal Register data on their patents Information on government assignees and government interest statements for these patents Data from NIH clinical trials database and FDA on whether NIH trials supported FDA approval	Government supported clinical trials for 6.6 percent of the drugs Government owned or supported patents for 7.2 percent of the drugs

Authors	Question	Empirical Approach	Measures	Results
Stevens et al (2011)	On how many drugs and vaccines emanate from public sector research institutions?	Examine number of drug approvals in-licensed from PSRIs (excluding licenses to platform technologies)	FDA data on drug and biologic approvals Orange Book data on FDA approved drugs AUTM data on academic patents and licenses rDNA data on licensing transactions	153 FDA-approved drugs discovered by public sector institutions over past 40 years (102 NMEs, 36 biologics, 15 vaccines) 13 percent of NMEs (21 percent of priority NMEs) licensed from public sector research Virtually all important vaccines introduced over past 25 years come from public sector Broad correlation between NIH Institute budgets and therapy classes with public sector drugs
Kneller (2010)	How important are new companies/universities (and other actors) in drug discovery?	Examine place of employment of inventors on key patents for drugs	252 FDA approved drugs 1998-2007 Data on patents from Orange Book, Merck Index, other sources Data from concurrent publications and from interviews on inventors' places of	Overall 24% of drugs from universities By novelty: 31% of most scientifically novel drugs By priority: 30% of priority-review drugs

Authors	Question	Empirical Approach	Measures	Results
Morlacchi and Nelson (2011)	What were the sources of innovation behind development of the left-ventricular assist device (LVAD)? How important was the NIH?	Longitudinal case study of the development of the LVAD	Interview data Information from key patents and publications on LVAD	NHLBI contracts important in spurring firm formation and evolution in 1960s/1970s NHLBI important in sponsoring conferences, centers to promote diffusion of best practice among academics and industry Public funding of key trials and development of component technologies also important Application led scientific understanding; basic understanding of heart failure remains weak

employment

| Dorsey et al (2009) | Are new drug approvals by therapeutic area associated with NIH funding in those areas? | Correlations of NIH funding data with future drug approvals | 1995-2000 FDA drug and approvals, mapped to nine disease areas NIH funding by Institute; allocated to disease areas based on Congressional justifications Note: Also estimate R and D by biotechnology firms, medical device firms, pharmaceutical companies, non-profits | Despite a rise in NIH (and other funding), drug approvals flat overall Within class analyses of drug approvals also show little correlation with research inputs |
| Blume-Kohut (2009) | How does NIH funding in a disease area relate to the number of drugs subsequently in Phase I and Phase III trials in that area? | Panel regression | CRISP and RePORTER data on NIH grants/funds 1975-2004 Grants associated with disease areas using parsing of abstracts, keywords, concordance with MeSH thesaurus PharmaProjects data on drugs in development, by phase and category | Some evidence of responsiveness of Phase I trials: elasticity .25-.31 No evidence of responsiveness of Phase III trials |

TABLE D-3 Public Funding and Private R and D, Patenting: Summary of Selected Studies

Authors	Question	Empirical Approach	Measures	Results
Ward and Dranove (1995)	How does industry funded R and D respond to NIH R and D?	Panel regressions relating private R and D in a disease area to NIH R and D by relevant institute	PhRMA data on R and D by field NIH data on R and D by institute Controls for disease burden, drug development, time	A 1 percent increase in NIH research associated with .76 percent increase by private sector over next seven years (direct) A 1 percent increase in NIH research associated with 1.7 percent increase by private sector over next seven years (indirect) Contemporaneous correlations highest
Cockburn and Henderson (1996)	How does interaction with public sector science (collaboration, hiring of "star" scientists) affect firm-level R and D productivity	Panel regression models relating productivity to within firm variation in interaction with public sector, with firm fixed effects	MEDLINE data from 35,000 articles on firms' co-authorship, publication by "star" scientists for 10 firms, 1980-1988 Data on "important" patents/R and D for these firms	Statistically significant association between propensity to co-author with academics and important patents/dollar Statistically significant association between share of publications from "star" scientists and important patents/R and D dollar
Toole (2007)	Does public scientific research complement private	Panel regression models relating pharmaceutical R and D by to NIH	CRISP data on NIH basic and clinical research mapped to 7	Public and private sector research complements A 1 percent increase in basic

Authors	Question	Empirical Approach	Measures	Results
	R and D investment?	funding across disease areas, over time	therapeutic classes, 1972-1996 PhRMA data on private sector R and D in these classes, 1980-1999	research funding associated with a 1.7 percent increase in private sector R and D A 1 percent increase in clinical research funding associated with a .40 percent increase in private sector R and D
Azoulay, Graff Zivin, Sampat (2011)	Do elite life scientists benefit local firms?	Panel regression models examining geography of citations to scientists' work before and after they move	Data on 10,450 elite life science researchers (most publicly funded) Historical information on productivity, employment locations of each scientist MEDLINE data on their publications ISI data on citations to their publications USPTO data on their patents USPTO data citations to their patents and publications	Professional transitions lead to a decrease in citations (in patents and articles) to movers' pre-move patents at original location Weaker evidence of increase in citations from firms at destination location
Zucker, Darby and Brewer	How important was academic science in the creation of new	Panel regression models relating location of new biotechnology firms to	337 "star" scientists (based on articles, genetic discoveries in	Presence of stars and their collaborators – "intellectual capital" – in an area has a

Authors	Question	Empirical Approach	Measures	Results
(1998)	biotech firms?	number of "star" scientists in area	Genbank) Data on their collaborators Location and affiliation of stars (from journal articles Data on biotechnology firms and firm formation form North Carolina Biotechnology Center and Bioscan	statistically significant and positive relationship with the number of new biotechnology enterprises later formed in that area
Cohen, Nelson, Walsh (2002)	What are the roles of public sector research on industrial R and D? What are the channels through which public research affect industrial R and D?	Survey	1994 Carnegie Mellon Survey of Industrial R and D managers Merged with publicly available data on respondents	Pharmaceutical industry an outlier: reports public research the most important source of new project ideas *and* contributing to project completion Medical instruments industry R and D projects less frequently use any of three outputs of public research than other industries Drug industry makes use of public research much more frequently Top three fields contributing to R and D in pharmaceuticals: Medicine, Biology, Chemistry Top three fields contributing to R

Authors	Question	Empirical Approach	Measures	Results
				and D in medical instruments industry: Medicine, Materials Science, Biology
Mansfield (1998)	How important is academic work for industrial innovation?	Survey	Survey results from 77 firms	Percent of new products that could not have been developed (without substantial delay) in absence of recent academic research, 1986-1994: 31 in drugs/medical products (15 across all industries) Percent of new processes that could not have been developed (without substantial delay) in absence of recent academic research, 1986-1994: 11 in drugs/medical products (11 across all industries)